How Physics Confronts Reality
Einstein was Correct, but Bohr Won the Game

How Physics Confronts Reality

Einstein was Correct, but Bohr Won the Game

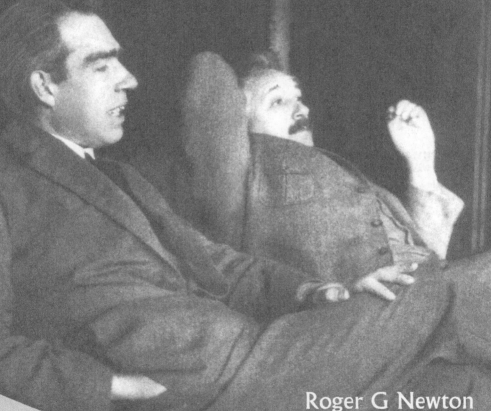

Roger G Newton

Indiana University, USA

World Scientific

NEW JERSEY · LONDON · SINGAPORE · BEIJING · SHANGHAI · HONG KONG · TAIPEI · CHENNAI

Published by

World Scientific Publishing Co. Pte. Ltd.

5 Toh Tuck Link, Singapore 596224

USA office: 27 Warren Street, Suite 401-402, Hackensack, NJ 07601

UK office: 57 Shelton Street, Covent Garden, London WC2H 9HE

British Library Cataloguing-in-Publication Data
A catalogue record for this book is available from the British Library.

HOW PHYSICS CONFRONTS REALITY
Einstein was Correct, but Bohr Won the Game

ISBN-13 978-981-4277-02-0
ISBN-10 981-4277-02-9
ISBN-13 978-981-4277-03-7 (pbk)
ISBN-10 981-4277-03-7 (pbk)

Printed in Singapore.

Preface

The early 20th century saw three radical changes that transformed theoretical physics and determined most of the course of physics ever since: the introduction of the theories of relativity, special and general, and of quantum theory. The last of these represented the really revolutionary upheaval. Albert Einstein originated the first two, and he and Niels Bohr initiated the third. However, after the full architecture of quantum theory had been designed by Werner Heisenberg, Erwin Schrödinger, Paul Dirac, and Max Born, the paths of the two revolutionaries who started it all diverged and they became scientific antagonists.

The fundamental point of disagreement between Einstein and Bohr was their clash concerning how quantum mechanics, as the fleshed-out quantum theory was called, dealt with the reality of the physical world. Both agreed that the new theory was based on describing and predicting the probabilities of the outcomes of experiments. Einstein insisted that therefore it did not describe reality. A theory that did not directly deal with reality and concerned itself only with probabilities of observations was unsatisfactory to him and needed completion. For Bohr, the outcomes of experiments were sufficient and, in fact, all that could be said about the sub-microscopic world.

This book recalls, for non-scientific readers, the history of quantum mechanics, the main points of its interpretation, and Einstein's objections to it, together with the responses engendered by his arguments. We point out that most popular discussions of the strange aspects of quantum mechanics ignore the fundamental fact that Einstein was correct in his insistence that the theory does not directly describe reality. While that fact does not remove these counterintuitive features, it casts them into a different light.

In order to set the question of how physics treats reality in its proper perspective, we then follow the history of two central aspects of physics: the elucidation of the basic structure of the world made up of particles, and the explanation, as well as the prediction, of how objects move. This history, prior to quantum mechanics, reveals that whereas theories and discoveries concerning the *structure* of nature became increasingly realistic, the laws of motion, even as they became more powerful also became more and more abstract and remote from intuitive notions of reality. The long history of atomism, beginning in ancient Greece, very slowly evolved from purely philosophical speculation to a theory with some basis in scientific facts. But by the beginning of the 20th century, atoms were firmly recognized as real. When it comes to accounting for the dynamics of moving bodies, the story is quite different. Aristotle's law of motion described intuitively how heavy carts moved against the resistance of the road, but the much more universal and powerful laws established by Newton gained their abstract power by sacrificing direct and intuitive contact with real experience, as every student new to learning physics can attest. The break with a direct description of reality two hundred and fifty years later embodied in quantum mechanics was nevertheless profound.

Another area of theoretical physics in which, in the 19th century, imagination and model building replaced direct description of reality was that of electromagnetism. Faraday's introduction of the seminal concept of a field and Maxwell's mathematical characterization of this field, though based on ideas that were later discarded, turned out to be of crucial importance in physics throughout the 20th century.

This partial history is followed by a brief account of the enormous achievements of physics based on quantum mechanics and its combination with Faraday's notion of the field, the quantum theory of fields. The accomplishments of physics owed to quantum mechanics all concern details of the fundamental constitutional structure of the world rather than how objects move. The fact that the quantum theory is probabilistic and does not directly touch reality in its basic architecture has not prevented it from being able to account for structural details of nature, intimately tied to reality and experimentally verified to an astounding degree of accuracy, and to bring systematic order to a welter of apparently chaotic experimental discoveries of new particles. As a result, more than half a century after Einstein's death and almost as long after the death of Bohr, there is no doubt among physicists that quantum mechanics, reality-friendly in its basis or not, is here to stay. Physics enters the 21st century with a number of profound open questions, but the validity of quantum mechanics is not one of them.

Contents

Introduction

If the purpose of physics is to correctly describe nature, then two of the various aspects of this description are among the most important: what the constituents of the world are and how objects move. We want to know what the world consists of and what accounts for the changes we constantly observe, that is, what are the dynamical laws underlying these motions.

Any physical theory intended as an explanation of dynamics, of the changes occurring in the universe and the movements of the things in it, must of necessity be deterministic; otherwise it would not be regarded as explanatory. This does not mean that we have to assume the world necessarily to be causal in the sense that the occurrence of all events can be predicted on the basis of earlier events, but it means that the state of the universe today allows us to predict its state tomorrow and next year.[1] If nature were so constituted as to preclude such a deterministic description, the science of physics could not exist. This is part of what Einstein meant when he said

---

[1] Please note that I am using the word "causal" to mean that the occurrence of events is predictable on the basis of earlier events, and the word "deterministic" to mean that the state of a closed system, such as the universe, at one time determines its state at all later times. My use of these words differs from that of some other authors.

that "[t]he eternal mystery of the world is its comprehensibility… The fact that it is comprehensible is a miracle."[2]

The miracle Einstein was referring to, however, requires some help from us. In order for a dynamical theory to allow the future state of the universe to be determined by its state at the present time, the theory must define what is meant by the "state" of the universe appropriately. If the state of a system of particles were defined simply as the collection of their position coordinates, Newtonian mechanics would not be deterministic (although Aristotelian mechanics would be, had the ancient philosopher aspired to such ends). Using Newton's equations of motion, you cannot determine the future positions of the particles from their present positions alone. These equations are such that both the initial positions and the initial momenta (or velocities) of all the particles are required — and sufficient — for their future motions to be uniquely predictable. Therefore, the *state* of a system of n particles in classical mechanics is defined by a collection of $6n$ numbers: three numbers for each of their position coordinates and three numbers for each of their momentum (or velocity) coordinates. Once the state of a system is given, the values of all its other dynamical variables are fixed as well.

Thus, the classical state of a system of n particles can be thought of as a point in a $6n$-dimensional space called the phase space. One of Einstein's miracles of comprehensibility can then be pictured in classical mechanics as the fact that each closed system of particles follows a unique trajectory in its phase space.

As we shall see, there is an exact analog of this fact in quantum mechanics. However, in order for quantum mechanics to be a deterministic (not necessarily causal!) dynamical theory, the state of a system has to be defined probabilistically and with only an indirect connection to the reality of that system itself. This is precisely what lies at the bottom of the many weird aspects of quantum mechanics

---

[2] Einstein, *Ideas and Opinions*, p. 292.

and what drove Einstein to his opposition against it as a fundamental theory. The following chapters will recall this history as well as recount the earlier story of those parts of physics that deal with motion and structure. The long history of dynamics shows theories becoming increasingly abstract and losing intuitive contact with real experience, whereas the description of the world in terms of atoms became more and more real, although the atomic structure itself remained without explanation. As we shall see, it turned out that quantum mechanics was able to account extremely successfully for the structure of the world, which classical physics was quite unable to do. That the same theory could deal only probabilistically with dynamics has turned out to be of little consequence, all its strange and counterintuitive aspects not withstanding. Even though Einstein was correct in his characterization of quantum mechanics as not directly dealing with reality in its basic architecture, the theory was a resounding success because it was able to explain remarkably accurately the observed particulate structure of the world. It is this ability that makes quantum mechanics still the most fundamental theory of physics.

1

Some Quantum History

The last decade of the 19th century had been a time of astonishing experimental discoveries. Among them was the identification by J.J. Thomson of cathode rays (emitted in the now somewhat old-fashioned cathode ray tubes of computers and TV sets) as consisting of negatively charged particles named electrons, weighing about one two-thousandth as much as a hydrogen atom, and the serendipitous discovery by the French physicist Henri Becquerel of radioactivity. Nevertheless, at the dawn of the 20th century, physics appeared to some at a dead end. The widely admired Lord Kelvin and others had recently declared that all that remained for the physicists of the next century to do was to add a few significant figures after the decimal point to their data. His vision was certainly clouded and his declaration dead wrong.

The opening shot of the coming quantum revolution was inadvertently fired by a man who could not have been more conservative, both in his science and in life. When in the year 1900 he found the solution to a conundrum that had been puzzling his colleagues, Max Planck was a well-established, productive 42-year-old German physicist, and his seminal idea was meant to be no more than a mathematical trick, though he realized its importance. No one had been able to give a satisfactory explanation for the observed change in the distribution of colors emitted by a red hot piece of iron, a "black body" — one that absorbed all radiation and reflected none — as its temperature increased. The problem should surely not have been beyond the powers of the science of thermodynamics, but all attempts led to nonsensical results. The device employed by Planck was to assume that when radiation of a given frequency, i.e., color, was emitted or absorbed by the black body, it always had to be in the form of discrete amounts of energy equal to that frequency multiplied by a fixed constant he called h. The trick worked, the law he derived by means of it agreed well with observations, and the constant h, forever after called *Planck's constant*, would echo and reverberate throughout the physics to come.

The True Revolutionaries

The true revolutionaries, the firebrands willing to violate laws they knew to be regarded as sacrosanct, were Albert Einstein and Niels Bohr, two men whose ideas would dominate most of physics for the 20th century. And, even though they much admired each other, they ended up in fundamental disagreement about the meaning of the revolution they had been instrumental in fomenting. But before we get to this disagreement, we will have to understand the theory that their novel ideas spawned and the history of how it came to be.

As a 25-year-old clerk in the Swiss patent office in Bern, Einstein produced three seminal papers during the year 1905, but he regarded only one of them as revolutionary. This was the paper usually identified as explaining the photoelectric effect. Never mind the one whose experimental verification finally served to persuade everyone that molecules really existed, or the one introducing the theory of relativity, which would change our concept of space and time forever. The "photoelectric" paper was really about the nature of light. It also explained the puzzling features of the emission of electrons from the surface of an electrically charged metal struck by light, called the photoelectric effect, but that was just one of its relatively minor implications.

Figure 1. The fringes produced by light passing through a slit, as first observed by Thomas Young.

The principal idea of this revolutionary paper was that light consisted of particle-like entities he called "quanta." This did not mean that Einstein denied the validity of the experiments performed by Thomas Young a hundred years earlier, which had definitively established that light was a wave phenomenon.

The interference fringes Young had observed could not have been produced by Newtonian particles; only the alternating constructive and destructive superposition of waves could result in the light and dark bands seen on a screen after a light beam had passed through a slit. Light of different colors consisted of waves oscillating with different frequencies. Furthermore, according to Maxwell's monumental theory of electromagnetism, it was the electromagnetic field that did the oscillating: light was an electromagnetic wave. Einstein did not deny any of this, but claimed, nevertheless, that light also consisted of "quanta," each containing an amount of energy equal to the frequency of the light wave multiplied by Planck's constant h and a momentum whose magnitude was equal to that energy divided by the velocity of light. Planck's way of deriving the correct law of black-body radiation, Einstein insisted, was not a mere mathematical trick needed to account for emission and absorption, but this was the way light always behaved; it was part of its nature. How it could be both a wave and also consist of particle-like quanta remained mysterious. Einstein evaded the question by calling his suggestion "a heuristic point of view." But there it was: discrete packets of energy of light, later to be called photons.

Seven years later, the 27-year-old Niels Bohr arrived in Manchester, England, to begin an assistantship in the laboratory of Ernest

Rutherford. Ranked by some as the greatest experimental physicist of the 20th century, the New Zealand-born Rutherford was investigating the constitution of atoms, the once-thought smallest constituents of all matter. Ever since he had identified the beta rays emitted by radioactive elements as being made up of J.J. Thomson's electrons, the little particles were expected to be found inside atoms — it would later turn out that, though electrons were indeed inside atoms, these were not the ones making up the beta rays of radioactivity — but were they embedded like raisins in a positively charged dough that filled the atom, making it overall neutral, as Kelvin and Thomson imagined?[3]

A New View of the Atom

Just a couple of years before Bohr's arrival, Rutherford had discovered that the atom instead was almost all empty space, with its entire mass and positive charge concentrated in a central nucleus so tiny he called it "like a fly in a cathedral." He had found this out by having his assistants Hans Geiger (who later invented the Geiger counter) and Ernest Marsden bombard a gold foil with alpha particles — other rays emitted through radioactivity, which he and the chemists Frederick Soddy and William Ramsey had identified as doubly positively charged helium atoms[4] — with the result that some had been scattered by wide angles, a few even in the backward direction. This could never be explained unless the atom's mass and positive electric charge was confined in a hard kernel much smaller than the atom itself. So where were the little electrons, which were of course much too light to play any role in the scattering of the

---

[3] The atomic model envisaged by Kelvin and J.J. Thomson after the latter had discovered the electron replaced the model based on vortices in the ether that they had advocated earlier. For a history of the vortex theory of atoms, see H. Kragh, *The Vortex Atom*.

[4] See R. Reeves, *A Force of Nature*, pp. 68ff, for how Rutherford did this.

impinging, vastly heavier alpha particles? Well, Rutherford thought, they had to somehow circle about the nucleus, forming the atom's periphery.

The trouble was that, according to Maxwell's equations, such circling electrons would have to radiate light, thereby losing energy and spiraling toward the center. The atoms he envisaged would exist for only a very short period of time, until their electrons crashed into the nucleus. What is more, they would emit a continuous range of frequencies of light, rather than the observed very specific spectral lines that characterized each atom. These line spectra had been used by chemists for years like fingerprints to identify the presence of every element when heated. The simplest of them, the spectrum of hydrogen, had even been encapsulated by the Swiss school teacher

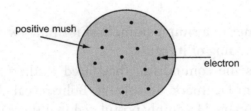

The atom of Kelvin and Thomson

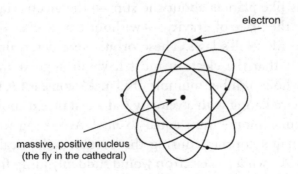

Rutherford's atom

Figure 2. The models of the atom proposed by Kelvin and Thomson and by Rutherford.

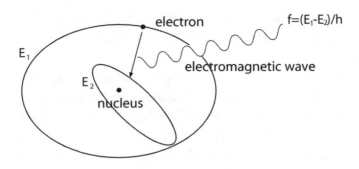

Bohr's hydrogen atom

Figure 3. The model of the hydrogen atom at two energy levels proposed by Bohr, including a transition from one level to another, accompanied by the emission of radiation.

Johann Balmer in a simple numerical formula (without, of course, any understanding of its cause).

This was the conundrum that faced Rutherford's young new assistant from Denmark. To solve it, Bohr revealed his true nature as a revolutionary. He simply postulated that there were certain specific orbits in which the electrons were allowed to circulate about the nucleus like planets about the sun — the electrostatic attraction taking the place of gravity — without emitting any radiation, contrary to Maxwell's laws. These orbits were determined by the requirement that the electron must have an angular momentum equal to a whole-number multiple of Planck's constant h. (A particle of mass m circulating with a velocity v at a distance d about a center has an angular momentum equal to mvd. According to Newton's laws, it remains constant during the particle's elliptical motion.) Every once in a while, an electron would suddenly jump from one of these otherwise stable orbits to one of lower energy, emitting the difference in energy $E_1 - E_2$ in the form of radiation with a frequency f given by Planck's $f = (E_1 - E_2)/h$. Only when an electron was in the orbit of lowest energy — he called it its ground state — was the configuration completely stable until the atom either collided with another atom or it absorbed light of just the right frequency to "excite" an electron to a higher level.

The model of an atom that Bohr postulated was totally unjustified by any physical laws or any coherent underlying theory. Moreover, Rutherford objected, when the electron started its descent and began to emit radiation, its eventual landing place would determine the frequency of the radiation it emitted. How did it know ahead of time on which level it would end up? Bohr did not answer such questions. For him the action of the jump and the emission of radiation with the correct frequency were actions that could not be further analyzed and he hoped eventually a comprehensive theory would be found that justified his daring postulates.

In the meantime, however, Bohr checked whether the spectrum of frequencies his model implied for the case of hydrogen agreed with the Balmer formula, of which he had just recently heard. Since the hydrogen atom was the lightest of all the atoms and thus had the lightest nucleus, it was reasonable to assume that it had the smallest positive central charge circled by a single electron. Its allowed orbits, according to Bohr's postulates, were easy to calculate and when he computed the differences between their energies, at least some of which should correspond to the series of spectral lines Balmer had neatly encapsulated in his formula, they indeed fitted. His model worked! Further confirmation arrived a couple of years later in an experiment performed by James Franck and Heinrich Hertz in Berlin. They sent a beam of electrons through a glass bulb filled with mercury vapor and found that when the electrons had certain specific energies, the vapor began to glow, emitting light of various spectral frequencies of mercury, depending upon the energies of the electron beam. These new results were easily explained by means of Bohr's model: when a beam electron of just the right energy collided with one of the electrons in the mercury atom, it would kick it up to a higher level, and the thus excited electron would subsequently descend to a lower level, emitting one of the spectral lines characteristic of mercury. Bohr's model must have obviously been right, even if it made no sense.

Both of the principal features of Bohr's atomic proposal were unexplained: the restriction of the electronic orbits to special, discrete energies where the electrons could remain at least for some length of time without radiating, and their spontaneous jumping from one orbit to one of lower energy, unless they were in their ground state. The first feature, the way Bohr had determined the stable orbits — stable, that is, except for occasional spontaneous emission of radiation — was much generalized and extended within the next ten years both by Arnold Sommerfeld in Munich and by Einstein. Bohr also pointed out that his atomic model, with many electronic orbits in heavier atoms, promised to lead to a much-needed physical basis for Mendeleyev's periodic table of the elements. Nevertheless, for a number of years the "old quantum theory," as it would later be called, remained *ad hoc* and could hardly be regarded as a full-fledged theory.

The second feature, leading to the question of how long an electron would remain in an excited state before jumping down, was elucidated about four years later by Einstein. Taking into account the presence of the ambient electromagnetic field surrounding every atom and interacting with the charged electrons, he calculated the probability per unit time for an atom in an excited state to descend spontaneously to a state of lower energy and emit the energy difference in the form of radiation, as well as the probability for it to either absorb or emit radiation when struck by a beam of light. (Einstein's novel idea that shining light on an excited atom could stimulate the emission of radiation, called induced emission, would much later turn out to be the basic principle of the laser.) The most noteworthy aspect of this paper is the fact that it was Einstein who for the first time introduced the notion of probability into the evolving quantum theory, the very aspect he later strongly disparaged. His so-called A- and B-coefficients, which determined these absorption and emission probabilities, have remained a permanent feature of quantum theory.

It should be noted that the two features of Bohr's model for the atom had quite different characteristics. One was structural, postulating the energy levels of the atom. The other pertained to dynamics and turned out to be probabilistic: an atom that was not in its ground state had a certain probability to change, and when it did, it emitted radiation. As we shall see, these two quite different features of the quantum theory followed two separate traditions in the history of physics: whereas, over the centuries, atomism slowly approached a realistic view of the structure of the world, theories of how objects moved became increasingly abstract and removed from real experience.

Einstein's light quanta, meanwhile, remained controversial for some years; even Bohr did not believe in them. However, in 1923 their existence was confirmed most directly by Arthur Holly Compton. Compton had found that when X-rays were scattered by electrons, the frequency of the rays scattered in any given direction differed from that of the incident beam exactly as you would expect the energy of a particle like the photon to change when colliding with a particle like the electron. Conservation of energy and momentum were the only laws needed to determine the result. The electromagnetic wave nature of X-rays could not account for the "Compton effect"; only their particle nature could.

In the meantime, the young Niels Bohr, the originator of a very successful picture of the atom, had become a famous man. However, it took until 1921, three years after the end of the Great War, for the Danish government finally to reward him, the first Danish professor of theoretical physics, with his own Institute for Theoretical Physics in Copenhagen, 44 years later to be renamed the Niels Bohr Institute. Einstein too had become famous, but not primarily because of his revolutionary work on quantum theory. He had become world renowned because of the General Theory of Relativity, a prediction of which had been verified in 1919 by the

British astronomer Arthur Eddington. Hailed as the successor of Isaac Newton, he had been appointed, with the strong support of Max Planck, as an especially privileged professor at the University of Berlin.

Constructing a Coherent Theory

The challenge of constructing a coherent theory to provide underpinnings for the *ad hoc* assumptions of the old quantum theory was taken up successfully by three physicists during the span of the three years 1925–1927. Two of them, Werner Heisenberg and Paul Dirac, were in their mid-20s, while the third, Erwin Schrödinger, was 38 years of age. A fourth, Wolfgang Pauli, who immediately tested Heisenberg's new theory by detailed calculations, was also just 25. No wonder it was dubbed *Knabenphysik*, or boys-physics.

After getting his PhD at the age of 22 under the direction of Arnold Sommerfeld, Heisenberg had served as assistant to Max Born in Göttingen, followed by an assistantship with Bohr. Inspired after listening to Bohr's lectures, he was imbued with the idea of constructing a grand new "mechanics" on the model of Newton but employing only directly measurable quantities as his tools. He fashioned his rules out of arrays made up of the "transition probabilities" Einstein had introduced in his A- and B-coefficients, but was puzzled when he found that these arrays obeyed strange multiplication rules: xy was not equal to yx. After Born recognized that his arrays were well known to mathematicians as matrices, which indeed made up an algebra with "non-commuting" multiplication laws, Heisenberg constructed what came to be known as "matrix mechanics." The equally young Viennese fellow student of Sommerfeld's, Wolfgang Pauli, immediately used Heisenberg's new theory to work out the possible energies of the hydrogen atom, and indeed they agreed with the *ad hoc* rules postulated by Bohr and led to Balmer's formula for the hydrogen spectrum.

Within two years, Heisenberg published what is probably the best known consequence of his theory, his famous uncertainty relation, which prevents the possibility of any arbitrarily precise simultaneous measurement of both the position and the momentum (or velocity) of a particle, as well as the simultaneous high-precision measurement of other pairs of "complementary" variables. Although the uncertainty relation could be derived mathematically on the basis of his theory, Heisenberg justified it by a physical argument that, in the case of position and velocity, went as follows.

If you want to find the position of a particle under a microscope as accurately as possible, you have to use light — or other electromagnetic radiation — of very short wavelength, such as gamma rays, in order to minimize diffraction effects that would blur its location. But the shorter the wavelength of the radiation, the more energetic its photons would be, and the harder they would kick the electron, as in the Compton effect, and therefore change its velocity in an uncontrollable way. In other words, he ascribed the origin of the uncertainty relation to an unavoidable disturbance during any measurement at the atomic level, an argument that would haunt the theory for many years. Widely misconstrued, it was transferred to other fields of science and even non-science without justification simply on the strength of its physics credentials. Although the uncertainty relation remained a permanent part of quantum theory, its origin from an inescapable measurement-induced disturbance was eventually shown by Einstein to be fallacious, as we shall see in Chapter 3.

While Heisenberg developed his matrix mechanics, another new idea was introduced into the quantum cauldron. Excited by Einstein's notion that light was constituted both of waves and particles, the French doctoral candidate Louis-Victor de Broglie had submitted a dissertation at the Sorbonne in which he proposed that, similarly, particles like electrons might, in turn, also have a wave nature. Specifically, he postulated that the wavelength l of the wave corresponding to a particle of momentum p should be $l = h/p$, where h is

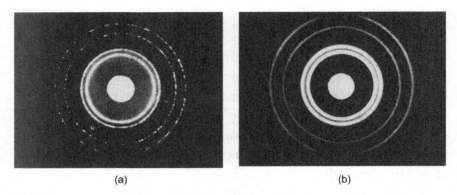

(a)   (b)

Figure 4. Diffraction patterns produced by sending (a) an X-ray beam and (b) an electron beam of the same de Broglie wavelength through a powdered crystalline target. (Picture (a) courtesy of the Education Development Center, Newton, MA; (b) courtesy of CAMECA Science & Metrology Solutions.)

Planck's constant. If that were the case, the electronic orbits in Bohr's atom might be intuitively justified as standing waves: the length of each orbit would be a whole-number multiple of the electron's wavelength and the wave would simply repeat itself as it went around. Consulted by de Broglie's professors at the Sorbonne about this weird idea, Bohr was lukewarm, but when Einstein was approached he was enthusiastic. De Broglie got his degree. What is more, within four years, the two American physicists Clinton Davisson and Lester Germer, as well as the Scottish physicist George Thomson, found that, indeed, a beam of electrons exhibited the same kind of interference effects that Young had discovered for light, thereby ascertaining its wave nature. If a beam of electrons is sent through a narrow slit in a screen and then made visible by allowing them to strike a second screen, the image will consist of several bright stripes rather than simply one bright image. These stripes can only be explained by interference. Particles though they were, electrons also behaved like waves. Everything in nature evidently exhibited a "wave-particle duality."

De Broglie's brainstorm not only turned out to be right, it also induced a brilliant new theoretical development. When Erwin

Schrödinger heard about it in Vienna, it stimulated him to invent an approach to the problem of transforming the *ad hoc* old quantum theory into a coherent theory that was entirely different from Heisenberg's. If particles behaved like waves, he would devise a suitable wave equation governing their propagation. The result was called "wave mechanics," with the Schrödinger equation at its center.

So in the span of three years two apparently entirely different theories, matrix mechanics and wave mechanics, had been spawned to undergird the old quantum theory. Of the two, Schrödinger's was certainly the more intuitive and attractive. Physicists were used to differential equations like the Schrödinger equation, whereas they were unfamiliar with matrices. How could they both be right? It did not take long for Schrödinger and others to show that, in fact, both were at bottom identical theories, expressed in different languages. The British physicist Paul Dirac, even younger than Heisenberg and Pauli, quickly reformulated the underlying theory in one common language, employing "operators" in place of matrices, and "bras" and "kets"[5] in place of Schrödinger's wave function.

How had Heisenberg and Schrödinger arrived at the equations that governed the way a quantum system behaved in the course of time? In the language in which Dirac had formulated their common theory, it was by starting with the classical Newtonian expression for the energy of a system of particles in terms of their positions and momenta, and if necessary, other "dynamical variables." All these variables would then be replaced by "operators," mathematical entities, for example a derivative, that change a given function into another. Such operators generally do not commute with one another, which means that pq is not equal to qp. In this manner the classical equations of motion are "quantized."

Finally, after the mathematician John von Neumann smoothed up its rough mathematical edges, "quantum mechanics" was in its mature form — except for the important question of what it meant.

---

[5] A *bra* and a *ket* together made up a *bracket*. That's the origin of these names.

2

Rules and Interpretations

The meaning of Heisenberg's matrices was clear enough: each entry in it was meant directly to represent a physical transition probability of the kind Einstein had used in his A- and B-coefficients. But what was the meaning of Schrödinger's wave function, the central concept of his theory, which satisfied his differential equation? Schrödinger's idea was that his ψ-function, as he called it, was analogous to an electric field. If the region where the ψ-function differed significantly from zero was not just a point but extended, then the electron itself was smeared out rather than being a point particle. This soon turned out to be untenable. For one thing his equation made the region where the ψ-function differed much from zero spread out very rapidly all over space, and for another, the ψ-function of three electrons was a function of nine coordinates, one for each of the three coordinates of the three electrons. It did not live in the three-dimensional physical space but in what physicists call "configuration space," which, for a system of n particles, has $3n$ dimensions.

It was Max Born who came up with the brilliant new interpretation that $\psi^2(x)$ (or more precisely $|\psi(x)|^2$, because ψ is a complex number) would designate the probability density of finding the particle at the position x, meaning that the probability of finding it in the small region dx was proportional to $\psi^2(x)\,dx$. Born's interpretation implied a profound alteration in the way the state of a physical system was described. Whereas in classical mechanics the state of a system of particles at a given time was described by specifying the positions and momenta of each of the particles at that time, in quantum mechanics that state was described by the probability of detecting them in certain positions, as well as the probabilities of the outcomes of other measurements. It made no reference to the positions or momenta of the particles themselves at all.

Both the classical and the quantum-mechanical definitions of state were designed so that the equations of motion made their development in the course of time deterministic: if the state was given precisely at one time, it was determined for all times in the future — in the classical case by Newton's equations, and in the case of quantum mechanics by Schrödinger's equation. There was, however, a profound difference between the two definitions: while the classical state description could be, and was, taken to be at least a partial description of the reality of the system itself, the quantum-mechanical state description does not refer to that reality at all; it merely describes probabilities of the results of measurements on the system. Thinking about the state of a system of particles classically, imagining it in terms of their positions and momenta at a given time, but being forced by quantum mechanics to describe it instead by its wave function, is a wrenching process that is the source of much confusion. The two state descriptions are truly incommensurable. In quantum mechanics the state of a system is an abstract mathematical construct having but an indirect relation to the reality of the system itself. This was a fundamental change that, as we shall see, would lead to the defection of Einstein from the theory he had greatly helped to found.

Most of the detailed postulates and rules underlying quantum mechanics are too mathematical to be described here. However, there were two specific rules that have led to particular interpretative controversies. One of these is usually referred to as "the measurement problem," or the "collapse of the wave function," and the other is known as "entanglement."

The Measurement Problem

When a measurement of the variable A on a system in the state ψ is performed and the result is found to be *a*, the state of the system is assumed in general to change discontinuously. The reason for this is not hard to see. If the statement "A has the value *a*" is to have any meaning, it should imply that if the measurement of A were to be repeated almost immediately, the result should again be *a*. So the state of the system right after the measurement has to be such as to assure this, whereas before the measurement the state assured the result *a* only with a certain probability. A wave function that has the property of definitely yielding *a* upon measurement of A is called an "eigenfunction" of the operator A that corresponds to A; it is denoted by ψ_a. This important assumption, that a measurement of A with the result *a* discontinuously changes the state of the system to a state described by ψ_a, is called the "projection postulate," because the state ψ is projected to the new state ψ_a. Another way of putting it is to say that the wave function instantaneously "collapses" to its new form ψ_a.

An immediate mathematical implication of the projection postulate is that the simultaneous measurement of two variables whose corresponding operators do not commute is generally impossible. This is because the set of eigenfunctions of such operators do not coincide. If the measurement of A is immediately followed by a measurement of B, and immediately after that A is again measured, the new result is not assured to coincide with the first, unless the operators A and B commute. In fact, one can prove that if two operators like **q** and **p** corresponding to the position *q* and momentum *p* of a particle are assumed to satisfy the commutation rule $qp - pq = ih/2\pi$ (where *i* is the imaginary number $\sqrt{(-1)}$ and *h* is Planck's constant), as they do, then the accuracies Δq and Δp with which these two variables can be measured simultaneously is necessarily limited so that the product $\Delta q \Delta p$ is greater than $h/4\pi$. This is precisely what Heisenberg's uncertainty principle says. If Δq is chosen to be very

small, so that the position of the particle is measured very accu-
rately, then Δp is forced to be larger than $h/(4\pi\Delta q)$, and *vice versa*.
The smaller one of these error limits is chosen, the larger the other
one has to be. In other words, the uncertainty principle is a direct
mathematical consequence of the projection postulate, together
with the commutation relations.

It is important to be quite clear about the meaning of
Heisenberg's uncertainty relation. It says that if the state of a particle is such that there is a high probability of finding it within the
distance Δq of the position q, then the size of the region Δp within
which its momentum p has a high probability of being found is at
least as large as $h/(4\pi\Delta q)$, and *vice versa*. The size of the region
within which the particle's position or momentum has a high probability of being found upon measurement is the uncertainty of its
position or momentum. The smaller this region is, the more precisely the state specifies its position or momentum, respectively. The
uncertainty principle says that the smaller one of them is, the larger
the other one has to be. It does not say that a very precise measurement of one variable necessarily rattles the other in an uncontrollable
manner.

One of the objections raised against the instantaneous collapse
of the wave function is that the latter usually extends and differs
from zero all over space. How can it instantly change everywhere
to a new form? But clearly this is based on a misinterpretation of
its meaning. It would be a valid objection if the wave function were
what Schrödinger originally intended, analogous to an electromagnetic field, or if it directly described reality so that its collapse
implied an instantaneous change of the system itself everywhere.
But once it is understood that the wave function is an abstract construct, that the space on which it lives is not our physical space but
the purely mathematical "configuration space," the objection loses
its force. What collapses is not a physical entity extending over
physical space; it is a purely mathematical entity in a mathematical space.

Another point raised is not so much an objection as an apparent puzzle. What is so special about measurement that gives it its apparently extraordinary power? Apart from the fact that it is performed by means of an instrument usually described in classical terms — with pointer readings, clocks, and meter sticks — what seems to be special is its registration by a human observer. (Bohr always insisted that the classical nature of the measuring instrument was an essential ingredient, but that by itself would not seem to lead to the collapse.) It is the awareness of the measuring result by a human consciousness that, to some commentators, is the essential component. This thought has been the origin of a long line of arguments persuading some physicists and philosophers that nature is subject to the influence of human consciousness.[6] However, this conclusion is a distortion of the meaning of the wave function, which is strictly that of a mathematical probability; ψ is not a description of the physical system itself. And, of course, probability always depends on conditions, and these are usually designated schematically as "knowledge." Laplace based all of his fundamental contributions to probability theory on the idea that probability judgments about the course of nature were needed only because we lacked full knowledge of all its details. In quantum mechanics, as Bohr would insist, probability was meant to be basic, and not founded on limited knowledge. Nevertheless the conditions underlying a given probability statement are still usually called knowledge, even when there is no knower present. (This situation is quite analogous to that of Einstein's Special Theory of Relativity, which relies crucially on an "observer at rest" and a "traveling observer," without thereby implying

---

[6] See, for example, the book *Quantum Enigma* by Rosenblum and Kuttner, which argues that human consciousness plays an important role in the way nature behaves. The authors of this book are, of course, not alone in making this argument, which is based on a fundamentally erroneous interpretation of quantum mechanics. Another example is Euan Squires's book, *The Mystery of the Quantum World*.

that nature depends on human consciousness.) So if a measurement of the variable A is performed on a given system and the result is "known" to be a, then the state of the system after the measurement is generally different from what it was earlier. The reason for the change of its state is not that the result of the measurement was registered in some human consciousness but that future measurements on a system that is certain to yield the value a when A is measured generally yield results that differ from those made on a system in a state for which this is not the case. Consciousness of anyone performing the measurement has nothing to do with this.

Beginning with a lecture at a meeting on Lake Como in 1927, Bohr gave a wide-ranging interpretation to the uncertainty relation. Calling it "complementarity," he raised it to a special case of a general philosophical principle that asserted that all knowledge was subject to a Yin-Yang-like limitation, in which two aspects of a truth could not be simultaneously ascertained. The more accurately we attempt to know one of them, the less we can know about the other. In physics, this was exemplified by the puzzling particle-wave duality, and in biology by the impossibility of simultaneously studying the nature of life and the detailed molecular composition of a living body, or simultaneously studying consciousness and the composition of the brain. Some philosophers and philosophically inclined scientists loved it; others were not impressed by Bohr's amateurish philosophizing.

Entanglement

The second quantum-mechanical peculiarity that has elicited much controversy originates from the so-called "superposition principle": the assumption that if ψ_1 and ψ_2 denote two possible states of a given system, then so does $\psi_1 + \psi_2$. For one thing, this principle seems to violate the rules of probability theory, because probabilities, you will recall, are to be proportional to ψ^2. According to ordinary probability theory, the probability for two independent events is the sum of the individual probabilities, but this rule seems to be violated in quantum mechanics because $\psi_1^2 + \psi_2^2$ is not the same as $(\psi_1 + \psi_2)^2$. The two differ by the cross term $2\psi_1\psi_2$, which can be positive or negative. (More precisely, since wave functions are complex numbers, the cross term depends on the relative phase of the two wave functions.) For many years this apparent violation of probability laws led to arguments that the rules of logic underlying classical probability theory had to be changed for quantum mechanics. However, as Dirac clearly recognized in his influential early book on quantum mechanics, the cause of the apparent conflict was a correlation between quantum-mechanical states of a system implied by the superposition principle that had no classical analog and was counterintuitive. This strange correlation between states, one of the weirdest aspects of quantum mechanics, and the feature most frequently discussed in the popular literature, is exactly what Schrödinger would, in a famous paper of 1935, call *Verschränkung*, or, as it was translated, *entanglement*.

Entanglement between states of a system appears to be particularly strange when two subsystems, after interacting when in close proximity, separate and are too far apart to be able to influence one another. For example, when two particles are created in a state in which their total momentum has a fixed value and they separate, moving far apart from one another, then their individual states remain correlated. As a consequence, the result of a measurement of the momentum of one of them allows an inference of what the result of a measurement of the momentum of the other one would be without ever performing that measurement. The states of the two particles remain entangled, no matter how far apart they may be, until they interact with other systems. This is the situation that Einstein explored, together with two young collaborators, Boris Podolsky and Nathan Rosen, in a very influential paper published in 1935, which is usually referred to as EPR. By that time Einstein had become totally disenchanted with quantum mechanics as a fundamental theory, and the EPR paper was meant to show why.

3

Einstein's Defection

Einstein's alienation from the theory to whose beginnings he had contributed in essential ways, both by his invention of light quanta and the calculation of his A- and B-coefficients, began with Heisenberg's uncertainty principle. Heisenberg had argued that since a particle's position and momentum could not be determined simultaneously with unlimited accuracy, its future trajectory could never be precisely predicted. The appearance of probabilities was thus not just a matter of limited knowledge but it was an indication that, according to quantum mechanics, at the atomic level the rule of causality was at an end. Einstein was unwilling to accept this categorical judgment. For several years after the publication of Heisenberg's paper, he took every opportunity at scientific meetings attended by both Bohr and himself to use ingenious thought experiments to show how he could measure both the position and the momentum of a particle as accurately as the measuring apparatus allowed. Contrary to Heisenberg's original argument involving his "gamma-ray microscope," intended to show how the precise measurement of one variable necessarily uncontrollably disturbed the other, he attempted to find ways of measuring both with unlimited accuracy. At each of these meetings, Bohr, who had become the grand old man watching over the proper interpretation of quantum mechanics — he, Heisenberg, and Born had together hammered out what came to be known as the Copenhagen interpretation of the theory — would agonize over Einstein's cleverly thought-out fictitious experimental arrangements until the next day, and in the morning invariably triumphantly demonstrate their fatal flaws. To do so gave him no pleasure because he enormously admired and liked Einstein. However, he was convinced the theory was correct and felt it incumbent upon himself to rescue it. After several such encounters, Einstein gave up trying to shoot down the uncertainty principle. But he did not drop his objections to quantum mechanics.

One of his criticisms was his widely known refusal to replace strict causality by probabilities. His statement that he did not believe God played dice with the world is, of course, well known. But this was not really his most fundamental reason for regarding quantum mechanics as no more than a provisional theory, even though he could not deny its many successes in precisely accounting for atomic spectra and, eventually, many other facts of atomic structure. Its inadequate dealing with reality made it unavoidable, in his view, that a more fundamental theory must take its place. The insistence by Bohr and Heisenberg that only experiments could determine what had meaning and what was real — if the position of an electron could not be determined when its momentum was measured, they insisted, it simply had no position — he derided as "sterile positivism." In his early days, Einstein had been influenced by positivism, but in his later years he came to reject it.

It was not that Einstein had become senile and unable to follow the new ideas (he was 49 in 1927), as is sometimes more or less openly implied. But he was no longer the revolutionary he had been in 1905, and he clung to what he regarded as the fundamental necessity for physics to mirror reality, even at the atomic level. As late as 1924 he had made an important contribution to quantum mechanics. When the Indian physicist Satyendranath Bose sent him a manuscript in which he developed the new statistics appropriate for particles, like photons, that were fundamentally indistinguishable, he enthusiastically elaborated upon it, creating what came to be called Bose–Einstein statistics and predicting the phenomenon known as Bose–Einstein condensation, whose existence was experimentally verified many years later.

In order to see why probabilities and statistics for objects that are indistinguishable have to be different from those for distinguishable ones, think of the number of ways two particles "a" and "b" can be put into three boxes. There are three possibilities for the first and three for the second, making it a total of nine: (a, b, 0), (a, 0, b), (b, a, 0), (0, a, b), (b, 0, a), (0, b, a), (ab, 0, 0), (0, ab, 0), (0, 0, ab); but if "a" and "b" are indistinguishable, then the third is the same as the first, the fifth is the same as the second, and the sixth is the same as the fourth. Therefore you need to pay attention only to the different number of ways in which the three boxes can be filled with two particles: (2)(0)(0), (0)(2)(0), (0)(0)(2), (1)(1)(0), (1)(0)(1), (0)(1)(1); so there are only six possibilities.

The EPR Paper

In 1935, after Hitler had driven him out of Germany to his new home at the Institute for Advanced Study in Princeton, New Jersey, Einstein, together with two associates, Boris Podolsky and Nathan Rosen, published the EPR paper mentioned at the end of the last chapter. Its purpose was to use the quantum phenomenon of entanglement to demonstrate that quantum mechanics was not a complete description of what Einstein regarded as reality.

In the somewhat clearer form in which it has been presented by the American physicist David Bohm, the EPR argument makes use of the intrinsic angular momentum of the electron, called spin. Otto Stern and Walther Gerlach had discovered that a measurement of the projection of this spin in any given direction could have only two possible results: it was either $+h/4\pi$ or $-h/4\pi$; no intermediate values were ever found. Furthermore, its projections on two different directions could not be ascertained simultaneously, because the corresponding operators did not commute: if its vertical spin component was found to be $+h/4\pi$, so that after the measurement the particle was in a state of vertical spin component $+h/4\pi$, a subsequent measurement of its horizontal spin component had a 50/50 chance of yielding the result $+h/4\pi$ and a 50/50 chance that it will be found to be $-h/4\pi$. If the electron was in a state of vertical spin component up, it could not at the same time be in a state of definite horizontal spin component east. According to the Copenhagen interpretation, this meant that an electron which definitely had a vertical spin projection pointing up did not possess a definite horizontal spin projection.

The Bohm version of the EPR argument then goes as follows. Suppose a particle of zero spin (that is, of zero angular momentum) decays into two electrons, which fly off in opposite directions. Since angular momentum is conserved, the total angular momentum of the two electrons must still be zero, i.e., they must be in a state in which their spins add up to zero. At a time when they are far apart and can no longer influence one another, we measure the vertical spin component of the first and find the result that it is up. We can then conclude with certainty, without measuring it directly, that the second must be in a state in which the vertical spin component is down: if its vertical spin component *were* measured, it would necessarily come out with the result "down." On the other hand, had we measured the horizontal spin component of the first with the result "east" then we would know with equal certainty, without actually having to measure it, that the horizontal spin component of the second was "west." From this, EPR conclude that both the vertical and the horizontal spin components of the second particle are "elements of reality." Here is what they say: "If, without in any way disturbing a system, we can predict with certainty the value of a physical quantity, then there exists an element of physical reality corresponding to this physical quantity." However, quantum mechanics says that the electron cannot be in a state in which these two quantities had — and could be measured to have — definite values! Therefore, quantum mechanics cannot be a complete description of reality. As Einstein put it: "I am inclined to believe that the description of quantum mechanics has to be regarded as an incomplete and indirect description of reality, to be replaced at some later date by a more complete and direct one."

Bohr answered the EPR paper almost immediately by publishing a rebuttal that many physicists did not really understand — Bohr's manner of expressing himself tended to be Delphic — but that was nevertheless regarded by most as conclusive. The gist of it was contained in the statement: "The extent to which an unambiguous meaning can be attributed to such an expression as 'physical reality' cannot of course be deduced from *a priori* philosophical conceptions but must be founded on a direct appeal to experiments and measurements.... In fact, this new feature of natural philosophy means a radical revision of our attitude as regards physical reality."

Here you have the nub of the disagreement between the two founders of quantum theory: can our ideas of what constitutes physical reality, based as they are on experience and perceptions at the macroscopic level of the everyday world, necessarily be carried over into the submicroscopic atomic realm? Einstein takes it for granted that they can, even though the quantum-mechanical definition of such a system's state makes no reference to this reality; Bohr says they cannot.

Apart from the unambiguous exhibition of Einstein's objection to quantum theory, the EPR thought experiment makes one thing perfectly clear: the origin of Heisenberg's uncertainty principle is not to be found in a disturbance caused by a measurement. Heisenberg's "gamma-ray microscope" thought experiment and other analogous arguments designed to make the uncertainty principle intuitively understandable are really beside the point. By taking advantage of the entanglement of the states of the two particles, the

EPR arrangement allows the measurement of any spin component of a distant second particle without ever coming near it, and similarly for its position or momentum (which was actually the original EPR version). Unless you believe in "spooky action at a distance," as Einstein called it, there could not have been any disturbance of that distant second particle by the performance of a measurement on the first particle. Nevertheless, two different spin components of that particle could not be determined at the same time: a measurement of its horizontal spin component after the indirect determination of its vertical component would have a 50/50 chance of yielding east and a 50/50 chance of yielding west, just as it would have been if its vertical spin component had been measured directly.

Schrödinger's Cat

Another aspect of entanglement was highlighted by Schrödinger in a brief paragraph of a long paper he wrote shortly after reading the EPR article with great sympathy. If the Copenhagen interpretation were to carry the day, he felt he was sorry ever to have contributed to quantum mechanics. The main point of his example was to show that whatever was strange about the consequences of the superposition principle was not confined to the microscopic world; it would have to manifest itself also in the macroworld, where it would become unacceptable. His little example would become famous as *Schrödinger's cat* and is being endlessly discussed to this day.

He imagines a cat enclosed in an airtight steel cage together with a "diabolical device." The latter consists of a radioactive atom that has a 50/50 chance of decaying within an hour, emitting an alpha particle that would be registered by a Geiger counter, which in turn would release a hammer that would shatter a glass bulb filled with cyanide gas, killing the cat. After an hour, he says, according to quantum mechanics the cat is in a state that is a superposition

32

of being dead and alive; only upon opening the cage and inspecting it will the cat be definitely either dead or alive; this is clearly absurd.

There are a number of objections to this argument, but the most fundamental one is clearly that Schrödinger, along with everyone else parading this thought experiment, is ignoring that the quantum state is not a *description of reality itself*, that is, of the cat itself, either dead or alive, but it is only a description of the probability of finding it either dead or alive. The cat is no more half dead and half alive than would be a mountain climber scaling a dangerous peak with a 50/50 chance of falling off. Nevertheless, the very counter-intuitive quantum-mechanical superposition of states that is invoked in Schrödinger's thought experiment has been experimentally verified in research conducted more than sixty years later (without the use of a live animal!). A group of physicists at the National Institute of Standards and Technology, using laser technology not available in Schrödinger's lifetime, were able to produce a superposition of two trapped states of a beryllium atom, one with high probability in position A with its spin up and the other with high probability in position B, far away from A (by atomic standards) with its spin down. In a sense, the atom was both at A with spin up and at B with spin down. Only when its position was specifically tested would it reveal itself to be in one place or the other. The title of their paper was *A "Schrödinger Cat" Superposition State of an Atom*.[7]

Schrödinger's example demonstrates that Heisenberg's and Bohr's insistence that nothing has reality until it is observed or measured, if accepted at the atomic level, makes it necessary to find a convincing argument for the transition from the submicroscopic to the macroscopic world. We cannot really accept a theory that claims that the reality of whether the cat is dead or alive is somehow suspended until we open the cage to take a look.

---

[7] C. Monroe *et al.*, *Science* 272 (1996), pp. 1131–1136.

Finding such an argument took physicists many years, but they eventually succeeded, and we shall return to this point later on. On the other hand, once it is clear that even at the atomic level quantum mechanics says nothing at all about reality but confines its statements to probabilities of the outcomes of observations, we can deal even with the cat in the cage.

Ensembles

One way of meeting Einstein's objections was to fall back on an ensemble interpretation of the quantum-mechanical probabilities. Physicists had been used to that ever since the development of the kinetic theory of gases in the 19th century. This very successful theory explained all the observed properties of gases, including such a mysterious phenomenon as irreversibility, in terms of the statistical behavior of their enormously many constituents, each of which simply followed Newton's laws of motion. In this context, the notion of an ensemble had been introduced, which consisted of an imagined, infinite collection of identical systems, all starting out more or less the same way, and their statistical behavior in the collection would then determine the probabilities for individual systems. In the probability theory invented by Richard von Mises early in the 20th century the very definition of probability was based on such ideas; it was called the frequency theory of probability.

However, simply defining probability in this manner, as many physicists did, would not really suffice to answer Einstein. The implication of such a definition was that physics would never describe a single system but always ensembles, and to this he could not reconcile himself: "Is there really any physicist who believes that we will never get any insight into these important changes in the single systems, in their structure and their causal connections, regardless of the fact that these single events have been brought so close to us, thanks to the marvelous inventions of the Wilson chamber [see Chapter 4] and the Geiger counter [see Chapter 1]? To believe this is logically possible without contradiction; but it is so very contrary to my scientific instinct that I cannot forego the search for a more complete conception."[8]

To effectively implement Einstein's ideas, a small number of physicists thought, required the construction of a new theory of a netherworld of "hidden variables" that would underlie quantum mechanics analogous to the way the motions of unseen micro-scopic molecules explained the behavior of gases in the macroscopic world. Such a theory was indeed proposed by David Bohm and others in the 1950s. Very few physicists paid any atten-tion to it. The principal reason was that it merely promised to duplicate all the successes of quantum mechanics without any new results. Physicists were quite happy working with quantum mechanics, so why should they switch to a new theory? Had Bohm's theory come in the mid-1920s, before quantum mechanics and the Copenhagen interpretation took center stage, it might have had a chance to become dominant. By the time Bohm pro-posed it, it was too late. As far as Einstein was concerned, he never took it seriously either. During the last 25 years of his life, he was working feverishly on a new unified field theory that, he hoped, would supplant quantum mechanics. Bohm's hidden variables were a cheap way out, he thought, and he disdained the theory. His own attempts ended in failure.

There was another reason why many physicists ignored Bohm's substitute for quantum mechanics. The highly esteemed mathematician John von Neumann had proved a theorem that purported to show that it was impossible for a classical hidden-variable theory to duplicate all the results of quantum mechanics. If he was right, there had to be something wrong with Bohm's theory.

---

[8] Einstein, *Ideas and Opinions,* p. 318. ("Physics and Reality," from the Journal of the Franklin Institute, Vol. 221, no. 3, March, 1936.)

Enter John Bell

It was not until the 1960s that the Irish physicist John Bell discovered a flaw in von Neumann's argument: to prove his theorem von Neumann had made an assumption that rendered it inapplicable to Bohm's theory. The equations of motion governing his hidden variables included nonlocal "spooky action at a distance." What is more, Bell proved that if you wanted a substitute for quantum theory that achieved all its results, it could not both directly deal with reality and also be local. The apparent nonlocality implied by the entanglement of states[9] in quantum mechanics would necessarily turn into action at a distance in a "realistic" theory.

Perhaps more important, Bell managed to turn the fundamental conflict between the classical approach to physics and the quantum-mechanical approach from a philosophical question into one that could be decided by definite experiments. (Bell's ideas have been called "experimental metaphysics" by the science philosopher Abner Shimony.) He proposed simple numerical tests that definitely separated the measured values of correlations which could be accounted for by local, classical signals — that is, without spooky action at a distance — from those caused by quantum-mechanical means. Here is a schematic example of Bell's idea in the form of a thought experiment.[10]

Imagine a setup consisting of three boxes at some distance from one another (see Figure 5). The box in the center is a transmitter of signals to the other two, and each of the other two boxes contains a receiver of these signals as well as a red and a green light and a switch with two settings. Care is taken that there can be no communication to any of these boxes other than the specified signals.

---

[9] It should be noted that this apparent nonlocality cannot be used to send signals faster than light, which would be prohibited by Einstein's Special Theory of Relativity.

[10] This is essentially the same as the one in the author's book, *Thinking about Physics*, Princeton University Press, 2000, p. 170ff.

Receiver A

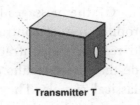
Transmitter T

Receiver B

Figure 5. An imaginary experiment illustrating Bell's idea.

The experiment consists of a number of runs as follows. At the beginning of every run the switch on each box is set at 1 or 2, the switch setting of box A being independent of that of box B. The transmitter then sends two uncorrelated signals, one to each box, and upon receipt of these signals each box flashes either its red or its green light, the color depending on both the signal received and the setting of its switch. The result of the run is recorded in the form, say, 2R1G, meaning: Box A, with switch set at 2, flashed red, and box B, with switch set at 1, flashed green. After many runs were completed, an examination of the recorded results revealed the following characteristics:

(a) neither 1G2G nor 2G1G ever occurred;
(b) 1R1R never happened;
(c) 2G2G did happen a number of times.

The problem now is accounting for these results in terms of classical signals, which we might denote by, say, [RG, GG], meaning signal to box A: if switch set at 1 flash red, if at 2 flash green; signal to box B: if switch set at 1 flash green, if set at 2 flash green. Let us then examine the recorded results. Characteristic (a) implies that neither [· G, G ·] nor [G ·, · G] was ever sent; (b) implies that the message [R ·, R ·] was never sent either. However, the result 2G2G, which did occur according to (c) could be brought about only by one of the four messages of the form [· G, · G]: the messages [GG, RG], [RG, GG], and [GG, GG] would produce results contradicting (a), and the message [RG, RG] would result in a contradiction with (b). We therefore have to conclude that these results could not have been produced by ordinary classical signals. They could, however, be accomplished by quantum-mechanical signals, which we shall not prove here.[11] In other words, in this thought experiment there is a definite procedure for deciding whether classical physics can account for the results or quantum mechanics is needed.

Bell's analysis did not remain at the level of thought experiments, even though real tests carrying out his ideas were not easy. Such experiments, employing linear polarization correlations of photons and using Bell's criteria to test whether a classical explanation would do or quantum mechanics was required, were carried out by the French physicist Alain Aspect and his group of collaborators in 1982.[12] Difficult though they were to do with the needed accuracy, they came out definitely in favor of quantum mechanics. Bell's contributions remain an active area of research in physics.

From Micro to Macro

There remained the question of why the strange quantum phenomena are not observable at the macroscopic level, in our everyday world. At one level the answer had been given early on by the Dutch physicist Paul Ehrenfest: he proved that the average quantum-mechanical values of a particle's coordinates satisfy Newton's equations of motion. Furthermore, it can be shown that, in general, particles behave classically in the limit as their de Broglie wavelength becomes very small compared to the distance over which their

---

[11] For a proof, see Newton, *op. cit.*, pp. 172–173.
[12] A. Aspect *et al.*, *Physical Review Letters* 49 (1982), pp. 1804–1807.

potential energy changes significantly (which can be looked at as the limit in which Planck's constant is allowed to become smaller and smaller). However, none of these arguments comes to grips with the much more serious mystery: why does the ordinary macroscopic world of our senses satisfy the familiar laws of probability, whereas in the atomic world the superposition principle reigns, which produces such strange phenomena as entanglement? Solving this problem turned out to be much more difficult and it took many years to find an answer.

The most satisfactory solution to the problem of the transition from the quantum laws of entanglement to the classical world of the usual rules of probability was developed by several physicists, and it goes by the name of "decoherence." The central question is why the cross term that is the difference between the sum of probabilities $\psi_1^2 + \psi_2^2$ and the superposition principle $(\psi_1 + \psi_2)^2$, namely $2\psi_1\psi_2$, which can be positive or negative, disappears. (More precisely, since ψ is a complex number, the cross term is proportional to the cosine of an angle that is the phase difference between ψ_1 and ψ_2.) A brief summary of the answer is that any realistic macroscopic system is in contact with an enormous number of small environmental influences, such as interactions with gas molecules and photons, and the effect of these interactions is to produce randomly oriented phase angles in the wave functions of interest, so that on average they will all wash out in their superpositions.[13] This is how the "coherent" superposition that is characteristic of the quantum-mechanical micro-world "decoheres" to the classical macro-world. This decoherence, together with the fact that macroscopic objects are made up of a huge number of sub-microscopic particles whose behavior, at the level of observation at which Planck's constant is negligibly small (which means that the uncertainties implied by Heisenberg's

---

[13] For a more detailed summary of the decoherence theory see the article "Decoherence and the Transition from Quantum to Classical" by Zurek, W.H., *Physics Today*, October 1991, pp. 36–44.

principle are negligible), is on average classical; it then also accounts for the fact that the laws governing the everyday world may be taken to describe reality. Even though quantum mechanics does not directly describe reality, and even though Bohr and Heisenberg may, or may not, be correct that at the atomic level physical reality comes into being only when a system is observed, the moon really is there when no one is looking at it. (Decoherence also frees Schrödinger's pitiable cat from being in a state of superposition, half alive and half dead, if it ever was in such a state. A large living system such as a cat could never remain in a superposition of states for more than an extremely short period of time even if it could ever be put into such a superposition.)

Was Einstein Right?

To sum up Einstein's negative view of quantum mechanics, I think it is fair to say that he was correct in his analysis that the theory does not directly describe reality, even though we may not necessarily agree with his definition of what constitutes reality. As any dynamical theory, quantum mechanics promulgates laws governing the temporal development of the state of a physical system, but its definition of the state of that system is an abstract mathematical construct that has no relation to a direct description of the system itself. On the other hand, it does not follow that he was correct in not accepting such a theory. One may well agree with Bohr when he declared, "[I]t is wrong to think that the task of physics is to find out how nature is. Physics concerns what we can say about nature," and concluded, "There is no quantum world. There is only an abstract quantum physical description." It also has to be said that Einstein's objections have had a very positive influence on the later development and clarification of the meaning and interpretation of quantum mechanics. It is no longer considered outrageous for a physicist to question the tenets of the Copenhagen "orthodoxy." What is more, without

Einstein it is unlikely that John Bell would have made his important contributions.

What we now have to do is to answer two questions that naturally arise at this point. First of all, is the distancing of quantum mechanics from a direct description of reality a totally new phenomenon in the history of physics, or has such a distancing been gradually taking place for a long time? Einstein recognized that it had been a historical process: "Physics constitutes a logical system of thought which is in a state of evolution… The justification (truth content) of the system rests in the verification of the derived propositions by sense experiences, whereby the relations of the latter to the former can only be comprehended intuitively. Evolution is proceeding in the direction of increasing simplicity of the logical basis. In order further to approach this goal, we must resign to the fact that the logical basis departs more and more from the facts of experience and that the path of our thought from the fundamental basis to those derived propositions, which correlate with sense experiences, becomes continually harder and longer."[14] Even more explicitly, he pointed to the electromagnetic field theory of Faraday and Maxwell as "a step in the direction of constructive speculation which has increased the distance between the foundation of the theory and sense experiences."[15] In the remainder of this book we shall address this question by examining the history of two parts of physics: first, its account of the constitution of the world in terms of particles or what used to be called atomism (Chapter 4), and second, the way physics accounted for the motions of objects, both in the heavens and on earth (Chapter 5). We shall see that whereas with respect to atoms and the particulate make-up of nature, physics has become increasingly realistic, its formulation of laws of motion has become more and more abstract, purchasing a vast increase in predictive power at the price of losing intuitive appeal and direct

---

[14] Einstein, *op. cit.*, p. 322.
[15] Einstein, *op. cit.*, p. 305.

contact with reality. We shall also flesh out, in Chapter 6, Einstein's statement about the electromagnetic field theory of Faraday and Maxwell as an example of "constructive speculation which has increased the distance between the foundation of the theory and sense experiences."

Following this we shall answer the question of what, in fact, were the great contributions of quantum mechanics by recounting the accomplishments of physics based on it, especially in its form as quantum field theory. While the substitution of probabilistic laws for classical causality has been a handicap in treating motion, the revolutionary new theory has been enormously successful in accounting for the structure of the world. Classical physics had been unable to do more than finding such structure and, in effect, acknowledging its existence without explaining why it exists, but quantum field theory has been able to account for it theoretically and to predict experimental results to an unprecedented degree of accuracy.

4

From Atomism to Real Particles

When we speak of the structure of the world, the most important issue is the question whether everything we see and touch is made up of individual particles or whether nature is a continuum. The idea that the world is made up of indivisible and unchangeable atoms goes back to Leucippus, a Greek philosopher who lived during the fifth century BCE, none of whose writings has survived. However, his student, Democritus, some 30 years his junior, became a famous philosopher and developed Leucippus's idea in greater detail. His fundamental view was that "Nothing exists except atoms and empty space. All else is opinion."

Democritus's atoms underlying the tumultuous changes apparent in the world we inhabit were tiny, hard, physically indivisible, indestructible particles existing forever as the immutable reality — quite unlike the ephemeral atoms of Hindu philosophy — and differing from one another only in their shapes. Though Democritus called them simple, some of these shapes were convoluted, consisting of various parts needed to account for the difference between solids, liquids, and gases. In a solid, the atoms had hooked branches that entangled one another and restricted their relative motions, whereas the atoms of a liquid were round and smooth, allowing more freedom of movement. Atoms had no color or taste and could generally move around freely, "like motes in a sunbeam on a windless day," jostling and bumping one another randomly without aim or purpose in empty space, the void that most subsequent philosophers regarded with distaste. Much as later thinkers continued to admire Democritus as a mathematician and philosopher, his indivisible atoms with variously shaped parts moving in a void always remained controversial.

Aristotle had no use for basic constituents of the world, too tiny to be seen, and his influence would later carry much weight with the anti-atomists during the Middle Ages. The Greek philosopher Epicurus, on the other hand, a contemporary of Aristotle though 40 years younger, accepted the atoms of Democritus, void and all, elaborating upon them and making them the foundation of his entire radically materialistic and anti-religious philosophy.

It was the Epicurean form of atomism that the Romans inherited from the Greeks, primarily through the voice of the poet Lucretius of the first century BCE. His great epic poem *De Rerum Natura* graphically sings of nature made up of an infinity of atoms roaming in an endless void.

Lucretius notwithstanding, as the Christian era approached, atomism lost its force. Both Cicero and Plutarch regarded the notion of naked atoms without qualities, rushing around in a void, underlying a world full of color, sound, taste, smell, and heat as manifestly absurd. Galen, the very influential physician who lived in the second century CE, would have none of them, and his prestige carried the day. Atoms were exiled from science and philosophy for the next 15 hundred years.

Atomism in the Middle Ages

Democritus was not entirely forgotten, but whenever his atoms were discussed, until the 17th century they were decisively rejected, often on the authority of Aristotle. As late as 1624, atomism was condemned by the University of Paris, and King Louis XIII of France threatened teachers of atomistics with the death penalty.[16] However, both the philosopher Francis Bacon and the scientist Robert Boyle accepted the concept of corpuscles, thus at least partially resurrecting atoms, but neither could bring himself to believe in the "hooked atoms" of Democritus, nor would René Descartes use them in his corpuscular worldview. The 17th-century French priest and philosopher Pierre Gassendi, an ardent follower of Galileo and an Epicurean who nevertheless rejected materialism, on the other hand, preached that atoms were the first things God created.

From this time on, when atoms had become acceptable again, they were generally identified simply with corpuscles and the original Democritean-Epicurean picture was for the most part ignored. Atomism began to shift from a purely philosophical concept to a gradually more scientific one, if not yet grounded on evidence; in a certain sense, the notion of atoms became more real. In contrast to the earlier, mostly metaphysical speculations, Francis Bacon favored an empirical approach to the theory of particles, and his attitude exerted a powerful influence upon Robert Boyle, who saw the world in mechanistic terms. Boyle also accepted both Bacon's view of heat as a vibratory motion of the smallest particles as well as René Descartes's theory that based the distinction between solids and fluids on the relative motions of these particles.

Boyle's great scientific influence rested primarily on his careful experimentation. The air pump he had constructed in his laboratory, with the assistance of the ingenious Robert Hooke, was able to achieve a far superior vacuum than any other device available elsewhere. But the very existence of a vacuum was a concept that philosophers such as Thomas Hobbes regarded as metaphysical and hence not subject to experimental test. The scientific notion of atoms moving in a void still had many enemies.

---

[16] Broda, *The Intellectual Quadrangle*, p. 5.

The "Corpuscular Philosophy" Boyle advanced included the idea that the basic particles of matter were for all intents and purposes impossible to cut or break apart, but he remained uncommitted and agnostic on the question of their fundamental indivisibility. However, he was a physicist and a chemist (though neither word existed at the time), not a philosophical systematizer.

Boyle played a decisive role in the transition from the mysticism and magic of alchemy to the newly developing science of chemistry, and his conviction that matter was made up of particles was based mostly on chemistry. Conversely, arguments in terms of particles always clarified his thinking about matters of chemistry. As Boyle was the scientist primarily responsible for reawakening atomism from its long sleep, it should then come as no surprise that when, some hundred years later, the existence of atoms was finally based on scientific grounds, the initial empirical evidence came almost entirely from the realm of chemistry.

Meanwhile the notion of atoms had to contend with the opposition of such scientist-philosophers as Gottfried Wilhelm Leibniz. Leibniz regarded the idea of atoms as the "smallest particles of matter" logically absurd. If they had extension, they could not be indivisible, and if they had no extension they could not be part of matter. Undeniably there was something logically incoherent in all the alleged attributes of atoms, but then, science is full of fundamental insights that start out incoherently expressed.

The First Scientific Revolution

Before the scientific confirmation of the philosophical speculation of particles as the ultimate constituents of matter occurred, the scientific revolution embodied in the work of Galileo Galilei and Isaac Newton had taken place. All its participants, in one form or another, believed in corpuscles; to believe in some kind of atoms became part of the newly prevailing anti-Aristotelianism. For Galileo's most significant work in physics and astronomy, particles played no important role, but he did express a strong belief in them. In his work, *Il Saggiatore* (The Assayer), he explicitly stated that sensations such as taste, sound, and smell could all be evoked by tiny particles without any qualities other than shape and motion. (Although attacked for such Epicurean notions by Church authorities, this issue did not play any noticeable part in his trial by the Inquisition.) Evidence that Newton accepted the atomistic philosophy is contained in his "Quaestiones quaedam Philosophicae," the notebook he kept in his youth, in which he indicated that he agreed with much of Gassendi's mechanistic philosophy and that he believed in atoms, but not necessarily in Gassendi's view of them.[17] In his *Optics*, he was quite explicit: "God in the Beginning form'd Matter in solid, massy, hard, impenetrable, movable Particles,... so hard, as never to wear or break in pieces..."[18] However, he later confused matters by at one time describing the particles of a gas as attracting each other, at another time as repelling one another, and later as independent hard spheres. As a result, Newton's authority had the effect of muddying the waters when it came to the constitution of gases. Notoriously, he also stuck with particles in his erroneous explanation of the nature of light.[19]

Christiaan Huygens, famous, among other things, for his wave theory of light, which opposed Newton's corpuscular theory, envisaged his light waves as oscillatory motions of the tiny particles he thought made up the luminiferous ether. So even Huygens's wave theory was based on the existence of particles. Meanwhile, the scientific evidence for the existence of atoms shifted, temporarily, to the realm of chemistry before returning, with full force, to physics.

---

[17] Westfall, *The Life of Isaac Newton,* p. 30.

[18] Kragh, *Entropic Creation,* p. 97.

[19] Newton's corpuscular theory of light, of course, was later partially resurrected in the form of Einstein's quanta of light, i.e., photons.

Chemistry had advanced since the time of Boyle and had been transformed into its modern form during the 18th century by Antoine Lavoisier, the great French scientist who lost his head under the Guillotine during the French revolution. Joseph Louis Proust had enunciated his law of constant proportions: every given chemical compound was formed from specific elements, and, in contrast to mixtures, which can be formed in arbitrary ratios, the ratios of the weights of these ingredients were always the same. This is where relevant chemical matters stood at the time of Dalton.

Atoms Based on Science

John Dalton, a mostly self-taught scientist — as a Quaker, he was ineligible for admission to the universities of Oxford and Cambridge — was a chemist, mathematician, and "natural philosopher." (The name "physicist" had not yet been coined.) Evolved from an initial fascination with meteorology, his interest in the properties of air had led him to learn that liquids like water were able to absorb gases. The only way he was able to understand how this could be was to conjecture that both gases and liquids must be made up of particles, so that the corpuscles of the gas could occupy the interstices between those of the liquid. This particulate view would carry him a long way.

The first result was his proposal in 1803 of what has become known as Dalton's law of partial pressures: the total pressure of a mixture of gases occupying a given volume equals the sum of the pressures that each of the gases in the mixture would exert if they occupied the same volume alone. His explanation was that the gases were made up of different kinds of particles which would exert repulsive forces on one another but would ignore the other kinds.

His interest turning to chemistry, Dalton generalized Proust's law of constant proportions to a law of multiple proportions: if the same elements could be combined to form more than one compound, their weight ratios in the different compounds would vary by factors that were small whole numbers. For example, the elements carbon and oxygen can be combined to form carbon monoxide or carbon dioxide. (These names for the gases, which imply their constitution, were, of course, not yet in use at the time.) A sample of carbon monoxide containing 1 g of carbon always contains 1.33 g of oxygen (the law of definite proportions), whereas a sample of carbon dioxide containing 1 g of carbon always contains 2.66 g of oxygen. The ratio of the weights of oxygen in the two compounds containing 1 g of carbon is therefore 1:2, in agreement with Dalton's law of multiple proportions.

The two laws he had discovered led him to the inescapable conclusion that the basic constituents of elements must be atoms (without any of the Democritean or Epicurean baggage), and those of compounds, molecules. (The word molecule for particles made up of more than one atom was coined by the French chemist Joseph Louis Gay-Lussac.) Moreover, he concluded from Proust's law that the ratio of the weights of the elementary gases forming a given compound must be equal to the ratio of the weights of the corresponding atoms. For example, the fact that carbon monoxide was formed by combining 1.33 g of oxygen with each gram of carbon indicated to him that each oxygen atom must weigh 1.33 times as much as a carbon atom. He was sometimes confused about molecules, not recognizing that an element could have constituents that are molecules made up of two or more atoms of the same kind. As a result he was wrong about some of his atomic weights, but his basic idea was profoundly correct. For the first time, the atomic concept had been put on a scientific foundation, and confirming evidence kept on accumulating for more than a century.

Working more or less in parallel with Dalton, though 12 years younger, Gay-Lussac had discovered two laws that would become very suggestive of a further refinement of Dalton's ideas. The first he announced in 1802: for a given rise in temperature, all gases expand by the same fraction of their volume. (But since the physicist

Jacques Charles had beaten him to it by five years, though without publishing it, it is now known as Charles's law.) The second, which might be regarded as a volume variant of the law of constant proportions, he established in 1808: when two gases combine chemically at a fixed pressure to form a new gas, the volumes they occupy always stand in simple numerical ratios. For example, two gallons of hydrogen and one gallon of oxygen combine to form two gallons of steam. As the molecules of the steam (made of water) contain two atoms of hydrogen and one atom of oxygen, this might suggest that the number of molecules in each gallon of these gases is the same. If that were the case, Gay-Lussac's law would follow as a simple consequence: the two gallons of hydrogen contribute twice as many hydrogen atoms as the one gallon of oxygen (taking into account that hydrogen as well as oxygen consists of diatomic molecules) to form a total number of water molecules equal to the number of hydrogen molecules and occupying the same volume, namely two gallons. Such a conclusion did occur to Dalton, but he rejected it. The reason was no doubt that if you envisage heavier molecules to be larger than lighter ones, and if you think of the atoms in a gas as closely packed together, which Dalton did, you will certainly resist the idea that a gallon of heavy molecules contains the same number as a gallon of light ones.

The conclusion Dalton unwisely rejected was courageously drawn by the Italian physicist Amadeo Avogadro on the basis of Charles's and Gay-Lussac's laws. He announced the hypothesis that at a given temperature, all gases contain the same number of molecules per unit volume. (This law, which has been experimentally verified in many different contexts, clearly implies that there must be relatively large distances between the molecules of a gas; they cannot be closely packed, as Dalton had thought.) A more convenient way of stating Avogadro's law is in terms of weight rather than volume, obviating the need to mention temperature: the number of molecules in a gram-mole (a quantity of gas whose weight in grams equals its molecular weight relative to the atomic weight of hydrogen) of any gas is a universal constant. The numerical value of this constant, now known as Avogadro's number, is approximately 6×10^{23}, but this was not established until the end of the 19th century; Avogadro had made no attempt at estimating it. About 10 years after Avogadro's death, the Austrian physicist Joseph Loschmidt was the first to make an estimate of its value; it is therefore also sometimes referred to as Loschmidt's number. He was even able to estimate the sizes of various molecules. The diameter of an average molecule of air was about one ten-millionth (10^{-7}) of a centimeter and the number of molecules in a cubic centimeter of a gas at room temperature was about 2×10^{18}. These numbers are not too far off presently accepted ones.

Persuasive as Dalton's evidence for the existence of atoms had been, it did not convince everyone. Whereas from the 19th century on chemistry could not do without atoms, much of the new physics, while often facilitated by models based on particles, did not require them. It was one thing for chemists to use atoms for intuitive purposes, but were they real? Thermodynamics, the physics of heat, was the arena in which the battle between those scientists who believed in the atomic constitution of matter and those who were skeptical of the reality of atoms was fought out most vociferously.

Atoms and the Nature of Heat

The 1769 invention of the condenser for the steam engine by the Scottish engineer James Watt made that engine, which utilized heat to perform mechanical work, the driving force of the industrial revolution. As a side effect of this economically and socially transforming development, the steam engine stimulated scientific questions about the conduction of heat and the behavior of materials, especially gases, under changing temperatures and pressures, bringing them to the forefront of science during the 19th century.

The crowning achievements of thermodynamics were the formulations of two fundamental laws of nature. The first was the law of conservation of energy, which governs the conversion of mechanical work and electrical energy into heat and *vice versa*. Based on the concept of a change of entropy (defined as the amount of heat flowing into a body divided by its temperature), the second law stated that the amount of entropy of any isolated physical system, such as the universe as a whole, can never decrease; it can only increase or remain the same. The implication of this law was that heat could flow only from a hot to a cold body, never the other way or even between two bodies of the same temperature.

While these purely phenomenological laws did not depend on any basic understanding of the nature of heat, which was still controversial at the time, they certainly cried out for such an understanding. What was the origin of these laws, and how were they related to other basic laws of nature? The second law was particularly puzzling because, for the first time, it introduced an arrow of time into physics: a process that increased the entropy of an isolated system could not be reversed, since decreasing the entropy was forbidden. It was clearly necessary to understand exactly what heat really was.

Plato had regarded fire — that is, heat — as one of the elements making up the world. But there had been no further progress in understanding what made a cold object warm up when in contact with a hot one, cooling the hot one down. Lucretius thought that heat was a substance flowing from a hot body to a colder one, and this intuitively very appealing idea was revived in the 18th century in a form that became known as the caloric theory: heat was an indestructible fluid called caloric, filling the interstices of all bodies, with the intrinsic property of flowing from a warmer object to a cooler one, like water flowing downhill. There was, however, also a quite different theory, which Robert Boyle had already accepted: what we sense as heat is nothing but an irregular motion of the particles making up a body. This came to be called the kinetic theory of heat.

Accumulating evidence gradually discredited the caloric theory, but it was the search for an underlying physical explanation of the laws of thermodynamics that finally established the kinetic theory, and with it, the existence of particles.

The "first law of thermodynamics," i.e., the law of conservation of energy, has a convoluted history, with multiple discoverers arriving at it along quite different paths. The road to discovery followed by the German physician Robert Mayer was circuitous and partly based on incorrect physics, but it led him to the conclusion that the conversion of work into heat was based on a fundamental conservation law of "force." (The word energy was not in general use yet.) He had made a profound discovery, but his arguments for it had been based on such faulty science that it took him several years of revisions to get it published.

Meanwhile, the British physicist John Prescott Joule, a man of exceptional experimental skills, had determined with great precision the "mechanical equivalent of heat," i.e., exactly how much mechanical work was required to raise the temperature of a kilogram of water by one degree. The publication of his work was generally ignored until it became the subject of an artificially whipped up priority dispute with Mayer, with nationalistic overtones that drew attention to both.

The concept of energy as something that was conserved in all mechanical processes — Joule subsequently proved it for electrical processes as well — therefore arose, together with the profound conservation law, out of quantitative demonstrations of its conversion from one form into another, specifically into heat. At the same time it solved the earlier mystery of why mechanical energy failed to be conserved in the presence of friction: the lost mechanical energy was simply converted into heat.

Now back to the question of what heat actually was. It was the German physicist Rudolf Clausius who delivered the quietus to the caloric theory and firmly established the kinetic theory in thermodynamics. He saw a gas as consisting of freely moving molecules, its temperature being a measure of their average "vis viva" (see Chapter 5). The enormous size of Avogadro's number made the cautious introduction of a statistical notion like average by Clausius necessary. There was no hope of keeping track of each individual particle. Whereas in an "ideal gas" there were no forces on the molecules except for an instant during collisions, in a liquid they were always close enough to exert forces on each other, and in a solid they vibrated and rotated about fixed centers.

An important objection to this final version of the kinetic theory of gases was that the average velocity of the gas molecules at room temperature was estimated to be of the order of about 2,000 feet per second. However, if a new odor was introduced into a room, it took a while to be noticeable at a distance. Clausius explained this by pointing out that because of the collisions among molecules, which deflected them from their original direction of travel, the molecules moved in zigzag paths that were much more extended than a straight line. Their diffusion from one end of a room to the other therefore would take a longer time.

Clausius's introduction of statistics and probability into kinetic theory, necessitated by the vast number of molecules in a gallon of gas, was completed by a trio of a Scott, an American, and an Austrian — James Clerk Maxwell, J. Willard Gibbs, and Ludwig Boltzmann — who created the new branch of physics called statistical mechanics (a word coined by Gibbs). In order to account for the properties of a gas, it was no longer sufficient to think of each molecule as moving with the average velocity of all of them, as Clausius had usually done. Such statistical details as their deviations from the average, and the way the velocities were distributed, etc., also played a role and were all worked out by Maxwell, Gibbs, and Boltzmann. But did this really mean that the reality of atoms was finally universally accepted? Even Gibbs still cautiously declared, "Certainly, one is building on an insecure foundation, who rests his work on hypotheses concerning the constitution of matter."[20] The man who dedicated his life to championing the reality of atoms was Ludwig Boltzmann.

The Role of Boltzmann

The great achievement of Boltzmann was to account for the laws of thermodynamics, and especially the mysterious second law, completely in terms of the behavior of the particles making up matter. He did this by relating the quantity called entropy, which Clausius had introduced on a purely phenomenological basis, to a measure of the probability of finding a specific arrangement of molecules. For example, if two rooms are connected by an open door, there are vastly more ways in which the positions and momenta of the molecules of the air in them can be arranged so as to be more or less uniformly distributed throughout the two rooms than there are ways in which the molecules in one room have a higher average kinetic energy than in the other. This is why the temperatures in the two rooms, initially different, become equal when the door is opened between them. The entropy of a system is a measure of the number of different arrangements of the molecules (their positions and momenta) in a given state of systems of the same kind with the same total energy; it is thus a measure of its probability. He even set up an equation governing the rate at which this probability changes with time. The second law of thermodynamics therefore simply says that a closed system, such as the universe as a whole, will evolve from a less likely to a more probable state. Since there are always many more states in which you would call a given system disordered than in which you would regard it as well ordered, the arrow of time implied by the second law leads from order to disorder. When a system

---

[20] In his book, *Elementary Principles in Statistical Mechanics Developed with Special Reference to the Rational Foundations of Thermodynamics.*

finally is as chaotic as it can be, it is in equilibrium and its entropy is at its maximum, unable to increase any further.

In this way Boltzmann managed to resolve the mystery of how the thermodynamic behavior of a system could show an arrow of time, i.e., behave irreversibly, even though it consisted of particles that followed Newton's laws of motion, which were reversible and contained no such arrow. This resolution had an implication that would revolutionize physics, with echoes throughout the 20th century. In contrast to what physicists working in thermodynamics had assumed all along, the second law, basic though it remained, no longer laid down an ironclad rule that the entropy of a closed system *could never* decrease; it merely said that the probability for it to decrease is extremely low. For many scientists this was a real shock. The hypothesis that matter was made up of a huge number of tiny particles turned out to imply that some kinds of behavior of matter were not predictable with certainty but only with high probability. For Max Planck, a specialist in thermodynamics, the shock was sufficient to make him temporarily doubt the atomic constitution of matter.

If I referred to atomistics still as a hypothesis, it was because Boltzmann had to spend his life fighting a constant battle against very reputable scientists in influential positions who still did not accept the reality of atoms: nobody had ever seen an atom! The quarrels blighted his life, especially since the hostility toward his work was concentrated primarily in his homeland and in the German speaking countries. At the age of 62 he committed suicide. His gravestone bears as an inscription the equation he had devised relating the entropy to the number of molecular states.

Who were the greatest of Boltzmann's scientific antagonists? His principal nemesis was the Austrian physicist and philosopher Ernst Mach, today popularly known for the eponymous speed of sound, named after him because of his work on shock waves. The basis of Mach's opposition to the reality of atoms was his advocacy

Figure 6.   Boltzmann's gravestone with the inscription S = k log W. (This image is from jacksofscience.com.)

of a philosophy of science founded on an extreme empiricism. As a Professor of Philosophy at the University of Vienna he taught the phenomenalist doctrine that reality consisted entirely of sensations and ideas and that sense impressions were the only things we could know of the world. Objects such as atoms were beyond our ken, and science had no business dealing with such "metaphysics." Mach's philosophical influence (sans opposition to atoms) on some scientists, including Einstein in his early days, would be felt long after his death in 1916. The question whether Mach ever accepted atoms before he died is somewhat controversial, but there is evidence that he finally did.[21] The tipping point appears to have been his viewing of the scintillations produced by alpha rays on a "spinthariscope," an instrument designed precisely for that purpose.

---

[21] Blackmore, p. 319ff.

Einstein Enters the Picture

As far as Boltzmann is concerned, there is a special poignancy in the fact that during the year prior to his tragic death, Einstein had published an article that would, for most scientists, serve as the final proof of the existence of atoms when its predictions were soon experimentally verified, and Boltzmann seems to have been unaware of it. (It is also possible that he consciously ignored it, since he had earlier specifically denied that the motion of molecules could cause observable movements of small particles.) The article by Einstein, one of the four extraordinary papers he produced during his *annus mirabilis*, 1905, explained a long-known phenomenon called Brownian motion.

In 1827, the botanist Robert Brown had observed under his microscope, as had other botanists before him, that plant pollens immersed in water appeared to execute a constant, irregular motion. Until 1905, this mysterious phenomenon was not understood, except that it was generally, vaguely believed to have something to do with the motion of the water molecules — if indeed they existed.

Einstein based his explanation of this random motion on the details of the fluctuations that would accompany the irregular movements of molecules as predicted by Boltzmann. These fluctuations would from time to time be so large that the resulting push exerted on a particle visible under a microscope would be unbalanced and move it on a path that would end up chaotic and zigzag. Einstein then proceeded to use known molecular masses and velocities at a given temperature to predict statistical details of this random motion, such as its mean velocity and the extent of its irregularity. The capstone of this visible confirmation of the existence of molecules was the experimental verification of Einstein's detailed predictions by the French physicist Jean Perrin.

For those still skeptical, the ultimately convincing proof of the existence of atoms was provided by some photographs presented in

1912 by the physicist C.T.R. Wilson, who had invented a clever device subsequently used for many years as a means for particle detection. Known as a cloud chamber, it was a glass container filled with cool, moist air, and closed with a piston. When the pressure was suddenly lowered by pulling out the piston at the same time as an electrically charged particle traversed it, the condensation trail of the particle became visible and could be photographed. The pictures Wilson presented showed the traces of alpha rays from a radioactive source he had sent through his cloud chamber. The most convincing pictures (see Fig. 7) — more convincing even than the scintillations in a spinthariscope — showed a trail that suddenly changed direction: clearly, an alpha particle had suffered a collision with a gas molecule in the chamber. The reality of atoms was finally, visibly established beyond doubt.

Let us, then, turn to a history of the laws of motion, and we shall see that in this area, the historical development was quite different. While classical physics was unable to account for the structure of the world, it was able to describe it, and this description gradually

Figure 7. Some cloud chamber pictures presented by C.T.R. Wilson.

became increasingly realistic, in fact so much so that some objected to a realistic portrayal of objects that no one had ever seen and would never see. In contrast to this, in the area of dynamics, that is, the accounting for motion, there was a distinct retreat from a direct description of reality in favor of intuition-defying abstraction.

5

Laws of Motion

Of all the observable motions in the world, those of the heavenly bodies were surely the most awe inspiring, and each of the ancient cultures, the Babylonian, the Egyptian, the Chinese, the Hindu, the Mayan, and perhaps some that left no records, made efforts to describe them. However, the philosophers of the Ionian civilization of the sixth century BCE, and Thales of Miletus in particular, were the first to try to understand what made nature tick rather than simply to describe it. Venerated as one of the seven legendary Wise Men, Thales left an indelible mark on the Greek civilization that germinated in Ionia.

In the fourth century, Aristotle took up the question of how to account for the observed changes in the world. These changes, he thought, including the many irregular ones on earth as well as the periodically recurring ones in the heavens, should all be subject to general laws. He divided the world into two realms: the translunar realm was the region of the majestic regularities of the stars, and there he would allow nothing but uniform motions along circles; the laws in the sublunar region had to be more subtle, taking into account the effort necessary to move heavy objects against the ever-present resistance offered by the ground and the surroundings. He therefore postulated that to move a mass M over a distance D

required an effort *F* exerted for a length of time *T* so that the product of the mass and the distance is proportional to the product of the effort and the time, with the constant of proportionality varying, depending on the resistance. Since the ratio of the distance to the time would be the average velocity *V* of the object, his law of motion for objects in the sublunar world was later interpreted to say that the product *MV* should be proportional to the force *F*, as in $MV = F/R$, where *R* is meant to be the resistance offered by the surroundings.

From a modern perspective it is easy to denigrate this Aristotelian law of motion, but in a practical world in which resistance to motion is ubiquitous, it makes perfectly good sense and agrees more or less with everyday experience when the resistance is strong. What is more, Aristotle recognized that it would not work for motion in a vacuum where there is no resistance. The existence of a vacuum, he concluded, was therefore impossible. Not until more than two millennia later would this perspective be reversed by Galileo's experiments, which minimized friction, and by Newton, who promulgated laws of motion that would hold both in the sublunar and in the translunar sphere. Since motion in the translunar sphere was subject to no resistance, his laws made the absence of resistance primary; for actual motion on the earth, where resistance is almost always present, it had to be appropriately modified. Until this change of perspective occurred, Aristotle's laws, with some minor suggested changes, were accepted. One of these changes was introduced in the 14th century by the mathematician and archbishop of Canterbury, Thomas Bradwardine. The flaw seen in Aristotle's law of motion was that, contrary to expectation, his law did not predict that when the applied force equaled the resistance, all motion would cease. He therefore postulated that in order for the velocity of an object to be doubled it would be necessary to square the ratio of the applied force to the resistance. This was a mathematical subtlety that accomplished its purpose without altering the underlying point of view. Ultimately, when disillusionment with

Aristotle's philosophy set in, his laws of motion were not the primary focus of the discontent.

This philosophy was part of the legacy the Greeks left to Western physical science. The changes in the world were assumed to be governed by Aristotle's separate laws for the sublunar and translunar spheres, and in the celestial realm they were fleshed out in the second century CE by the Egyptian astronomer Ptolemy. Living mostly in Alexandria, which by then had been conquered by the Romans, he designed a model for the motions of the planets that employed nothing but Aristotelian circular trajectories. However, in order to explain their retrograde motions — there were times when a planet seemed to be moving in a direction opposite to its normal motion — he had the centers of their individual circular orbits form epicycles moving along a larger circle centered on the earth. This artificial and complicated construction was meant to account for all the movements seen in the heavens and it seemed to satisfy astronomers, as well as the Church, for some 1,300 years, even though it did not really fit the observed data very well. In contrast to the laws governing changes in the sublunar world, where Aristotle demanded underlying causes such as exerted effort, the rules for the regular motions observed in the sky were not subject to causality but only to ideals of circular symmetry. The notion of reality did not play a significant role in that realm either; after all, no one could possibly go to the heavens and check up on what the stars actually were and how they moved.

The Great Break with Aristotle

The influence of Aristotle was still strong some 2,000 years after his death, and his accounts of motion, separately near the ground and in the heavens, was still in vogue. The scientific revolution engendered by Galileo and Newton was specifically centered on a break with this Aristotelian view. The first radical change in the celestial realm, introduced by Copernicus, which replaced the complicated geocentric model of epicycles devised by Ptolemy with a simpler heliocentric one, still retained Aristotelian circles as orbits of the planets, including the earth, all centered at the sun. However, Kepler's great improvement of the heliocentric model, with its elliptical orbits of the planets, was a direct insult to the Greek philosopher and his law of motion in the translunar area. While even Galileo disliked the elongated orbits for aesthetic reasons, there was no denying that Kepler's new laws were based on his observations rather than Platonic preferences.

The abrogation of Aristotle's laws was completed by Isaac Newton, who simply abolished the division of the world into sublunar and translunar realms. There was for him only one set of laws, valid everywhere. The falling apple was subject to the same gravitational force as the moon and the planets, and it reacted in the same way. Furthermore, as Galileo had already found, it was the acceleration A rather than the velocity V that was proportional to the applied force, and Aristotle's $MV = F/R$ was replaced by Newton's $F = MA$. In contrast to Aristotle, Newton idealized the everyday world by assuming there was no resistance, just as there was none in the sphere of the planets. Balls thrown were treated as though there was no air resistance, and we were all — in a first approximation — moving along slippery, icy streets.[22] Newton's laws were, of course, more likely to be applied to the motion of objects thrown through the air, often encountering relatively little resistance, than to that of carts on muddy roads. The new, vastly more abstract approach sacrificed intuitive realism but gained enormous mathematical and predictive power. (Even today, every teacher of elementary physics finds that most students are natural-born Aristotelians who find Newton's laws very counterintuitive and hard to learn.)

To account for the motions of the planets, Newton postulated that the force F was that of gravity, whose strength, he assumed, was

---

[22] The combination of Newton's laws, however, did allow the introduction of appropriate allowance for the effects of friction.

proportional to the product of the masses of the two bodies between which it acted and inversely proportional to the square of the distance between them. His laws of motion provided a framework in which the nature of the applicable force had to be supplied separately; the new universal law of gravitation supplied the most important one of these. The combination of the two allowed Newton to account for the motion of the moon and all the planets exactly in accordance with Kepler's laws. It worked for comets too. The astronomer Edmund Halley used Newton's laws to recognize that the comet he observed in 1682 was the same as the one seen in 1456 and described in detail by Paolo Toscanelli dal Pozzo; it had an elliptic orbit with a period of 76 years, he calculated, and he predicted that it would return in 1758. When on December 25, 1758, "Halley's comet" was indeed spectacularly observed in the sky, there could no longer be any doubt that Newton had mastered motion. What is more, the same laws valid in the sky served to predict — approximately — the trajectory of a ball shot from a cannon. The celestial sphere was no longer separated from that near the surface of the earth. However, even as the laws of motion on earth had become further removed from direct experience, so too had the universal law of gravitation become counterintuitive and abstract. Its action at a distance, which lacked any mediating mechanism transmitting the force of attraction between objects, thereby offending even Newton himself, took a long time to be accepted.

Problems of Motion Dominate Physics

From the 18th century on, then, physics began to be dominated by problems of motion, both in the heavens and on earth. At the same time, neo-Platonic structural notions, such as fitting the sizes of the planetary orbits into nests of "perfect" polydedra, as Kepler had attempted, were rejected. In Newtonian physics, the sizes of these orbits required no physical explanation; they were accidents of history.

Meanwhile, the physics of motion developed rapidly by employing Newton's innovations. In contrast to widespread initial opposition to Newton's law of universal gravitation, his laws of motion were generally accepted in quick order. In the course of the 18th century the mathematicians Jacob Bernoulli, Johann Bernoulli, Daniel Bernoulli, Jean d'Alembert, Leonard Euler, Joseph Louis Lagrange, and William Rowan Hamilton in the 19th century, as well as others, recast his laws into the form of a set of differential equations more amenable to general mathematical treatment no matter how many objects were involved.

These mathematically powerful forms of the Newtonian laws of motion led to the discovery that systems of interacting bodies contained, hidden among their ever-changing positions and velocities, certain *constants of the motion*. The motions, easily observable but usually very hard to calculate in detail, danced around a few elements of permanence. Abstractly defined though these elements had to be, their mathematical existence turned out to play an increasingly important role in physics.

One of these conserved quantities had first been used by Jean Buridan in the 14th century to argue against Aristotle's law of motion: the inertial impetus of an object, later named momentum and defined as its mass multiplied by its velocity, kept it moving even after all applied forces had ceased. However, it took about 60 years after the publication of Newton's *Principia* for Euler to discover that for any system of point particles in interaction with one another, the total sum of the individual momenta always remained constant.[23] This was called the law of conservation of momentum.

A second quantity that is permanently conserved for any system not acted upon from the outside is the total sum of *angular momenta*

---

[23] Each of the individual momenta has a direction as well as a magnitude, and they have to be added up *vectorially*, that is, taking their different directions into account.

with respect to fixed point P. If a point object of mass m circles about P at a distance d and speed v, its angular momentum is equal to the product dmv. The law of conservation of angular momentum comes into play, for example, when a twirling ice skater pulls in her arms from an extended position, thereby speeding up her twirling because, in order to keep her angular momentum the same, the decrease in the distance of her hands has to be compensated for by an increase in speed of rotation. Conservation of angular momentum is also responsible for the stability of a gyroscope; in this case it is the direction of the angular momentum, along the axis of rotation, that needs to be preserved.

The story of the third conservation law was a bit more complicated. It seems to have been Gottfried Wilhelm Leibniz who discovered that for mechanical systems consisting of colliding point particles with masses m_1, m_2,... and velocities v_1, v_2,..., the sum of all the products $m_1 v_1^2 + m_2 v_2^2 + \cdots$ always remained conserved. He called this sum the *vis viva* of the system, or the live force. Later, half that sum $(m_1 v_1^2 + m_2 v_2^2 + \cdots)/2$ was found more significant and called *kinetic energy*. However, when there are forces acting on the particles, for example, the force of gravity, this kinetic energy was not constant. The kinetic energy of a ball dropped from a tower continually increased as it accelerated. At rest at the top at height H, it had therefore the potential of acquiring energy of motion, kinetic energy, if released. So the concept of *potential energy* was introduced. Since the amount of work it took to transport the ball of weight w to the height H was wH (work being defined as the force multiplied by the distance over which that force had to be overcome), this potential energy E_{pot} of a ball of weight w at the height H was defined to be wH. Then the sum of its potential energy and its kinetic energy remained constant as it fell; as the ball drops, more and more of its potential energy is converted into kinetic energy. As the bob of a pendulum swings, there is a continual conversion of potential energy into kinetic energy as it descends and of kinetic energy back into potential energy as it rises.

Its total energy remains constant during the entire motion (in the absence of friction).

The concept of potential energy was then generalized to forces other than gravity, and mechanical systems containing only such "conservative forces" were subject to the law of conservation of energy. This, however, excluded all the systems with friction! Most practical mechanical systems did not conserve energy but lost some of it to friction. (Newton's idealization, after all, had ignored the resistance which Aristotle had built into his law of motion.) In the applications of Newton's laws of motion to celestial mechanics, i.e., the orbits of the moon and the planets, energy is conserved. But most moving objects on earth are subject to friction, and some energy gets lost. Every real pendulum eventually comes to rest and no ball will bounce forever. Until the 19th century, this appeared to be a bit of a mystery, whose solution we already discussed in Chapter 4.

The Newtonian revolution in physics had renewed and vastly strengthened the Aristotelian quest for the explication of all motion, both on earth and in the heavens. The work of 18th- and 19th-century mathematicians for the first time made it possible for the laws governing these motions to make actual long-term predictions of such motions, as we have seen. The French mathematician Pierre Laplace went so far as to announce that if only he had the necessary precise knowledge of all the particles in the universe now and if he were smart enough (in modern terms: if he had a powerful enough computer), he could predict its future course forever. However, since he also knew that no one could possibly have the needed precise knowledge of the initial positions and momenta of all the particles, he would have to make do with probabilities. So he proceeded to develop the calculus of probability, which later on turned out to be of profound importance.

At the end of the 19th century, the French mathematician Henri Poincaré would show how right Laplace had been in appreciating that the future course of the universe could not be predicted without the most precise knowledge of all the details of its present state. To his great surprise — and horror, since he had made a mistake in a prize-winning paper that he had just published — he found that even systems consisting of only three particles were "sensitive to initial conditions." This meant that the slightest error in the description of its present state would soon magnify itself to such an extent that its later state became utterly unpredictable: it, in fact, would become *chaotic*. (Since every modern computer is digital, its precision is limited by the finite number of digits in its computations. Poincaré's theorem therefore rules out any long-range computer computation of systems consisting of more than two particles.) So, although the motions of mechanical systems after Newton became, in principle, quite well understood, for practical purposes long-range prediction remained, for all but the simplest two-particle systems, unattainable. What is more, any notion that Newton's laws were a direct mirror of reality was clearly illusionary. These laws constituted a vast abstraction from an idealized concept of reality, and precisely this abstract quality was what gave these laws their power.

At the same time, the discovery of constants of the motion, or conservation laws, showed that underlying the welter of complicated movements of even large mechanical systems there were aspects of permanence. What was the explanation of the existence of such elements of repose in the midst of apparent, and sometimes real, chaos? That understanding did not come until the 20th century, and it injected an entirely new perspective into theoretical physics.

The New Perspective

The person primarily responsible for introducing this fruitful new perspective was a German mathematician named Emmy Noether. Her great achievement was to prove a theorem that connected two concepts which in all of subsequent physics would play a dominant role: conservation laws and symmetries. While, as we have seen, conservation laws had become important after the Newtonian revolution, the prominence of symmetry considerations was relatively new. Her theorem stated that whenever a theory is invariant with respect to a group of transformations, it necessarily contains a corresponding physical quantity that is conserved, i.e., does not change with time. Now, what does "invariant with respect to a transformation" have to do with symmetry?

When we say that a figure has "mirror symmetry" what we mean is that it does not change — it is invariant — when reflected in a mirror. Similarly, to have "rotational symmetry" means to be invariant when rotated. The concepts of invariance and symmetry are basically identical.

Suppose now that the equations of motion of a physical system are invariant with respect to shifts and reversal of time, just as Newton's equations of motion of many systems are. In other words, they are the same, whether the motion starts today or tomorrow or at any later time with the same initial conditions, and they remain the same when the direction of time is reversed. Then, Noether's theorem asserts, the energy of the system is necessarily conserved. Similarly, if the equations are invariant under a shift in space — it does not matter whether an experiment is performed here or a mile down the road — then the total momentum of the system is conserved; if they are invariant under rotations, then the total angular momentum is conserved. The conservation laws, Noether discovered, are consequences of the symmetries of the equations governing the motion of any physical system. The Platonic concept of symmetry was suddenly re-introduced into physics and acquired a new prominence. During the remainder of the 20th century, this prominence would rise enormously, as we shall see when we turn to the achievements of quantum mechanics and quantum field theory. But we must now return to further developments in the explanation of motion.

New Laws of Motion

During the same miraculous year of 1905 during which he had explained Brownian motion by means of molecules and the photo-electric effect by postulating the existence of light quanta, Einstein had introduced the Special Theory of Relativity. This theory implied a radical revision of the meaning of time and distance, a revision that was regarded by many as not only counterintuitive but totally nonsensical. To this day, physics journals from time to time receive submissions from non-physicists purporting to prove that Einstein's theory is illogical and cannot possibly describe reality.

An additional implication of this theory was that energy and mass where interconvertible (just as mechanical energy and heat had been found to be interchangeable in the 19th century) at the rate $E = mc^2$, where c stands for the speed of light. No longer was energy itself a conserved quantity, but only the totality of mass and energy. That no conversions from one to the other had ever been observed would soon change and dominate much of the physics of the 20th century. Concomitant with this acquisition of energy by virtue of its mass, a freely moving particle was postulated by the Special Theory of Relativity to possess a combination of kinetic energy and mass-energy. From a practical point of view, this change would be almost entirely unobservable for all motions that are slow compared to the enormously large speed of light, 300,000 km/s; it would matter only at speeds approaching that of light.

Twelve years later, Einstein enlarged the scope of his special theory to the General Theory of Relativity, a theory of gravitation that replaced Newton's law of universal gravitation and accounted for the motions of massive celestial bodies. Unlike Newton's theory, which consisted of two separate parts, the universal law of gravity together with the framework of his laws of motion, the General Theory of Relativity combined gravity directly with specifying motion. Its crucial ingredient was the fact, unaccounted for by Newton, that the same inertial mass that determined an object's resistance to acceleration also determined the strength of its response to the call of gravity. This equality of inertial and gravitational mass was the reason why, in vacuum, all objects on earth fall with the same acceleration.

The conclusion Einstein drew from this was that motion in response to gravity could be described in purely geometric terms. The needed geometry, however, would have to differ from the long-familiar one we owe to Euclid, the ancient Greek; the appropriate one was a "non-Euclidean" geometry invented in the second half of the 19th century by Bernhard Riemann. In this geometry, the local "curvature" (which fixes what the sum of the interior angles of a triangle should be, rather than Euclid's familiar 180°) is determined by the nearby masses, and all objects move along lines that are "straight." The path of a light ray from a star passing near the sun, for example, though straight according to Riemann, appears bent to us when judged by Euclidean standards. The detection of the exact size of this light-bending effect during a solar eclipse in 1919 was, in fact, the first observational evidence that Einstein's theory was correct. He was universally hailed as the man who had dethroned Newton. In addition, his theory was also able to explain the exact size of the rotation of the major axis of the planet Mercury's orbit, a long-known anomaly in Newton's theory.

Counterintuitive as the General Theory of Relativity was, at least to the general public, with its curved "straight lines" and triangles with interior angles not adding up to 180°, it was a testimony to Einstein's deep insight into reality as he perceived it. It was on the basis of his confidence that this insight could not mislead him in the subatomic world that he remained staunchly opposed to the later developments of quantum theory.

In contrast to quantum mechanics, general relativity successfully described motion, but at the submicroscopic scale, where the strength of the force of gravity was negligible, it played no significant role and could be ignored. For large-scale movements of stars and galaxies, on the other hand, it was indispensable, and the combination of the construction of new, much more powerful telescopes, the utilization of new means of observation by radio waves and neutrinos, together with the General Theory of Relativity led to a great expansion of our knowledge of the history of the universe. As a matter of principle, however, the so far irreconcilable conflict between quantum mechanics and general relativity constituted the greatest unsolved puzzle of 20th century physics. It was as though physics had returned to Aristotle's division between laws valid for the heavens and different laws down below; only the demarcation line was drawn differently. One part of physics successfully predicted large-scale motion, the other for the first time explained submicroscopic structure on the basis of general abstract postulates; and the theories employed did not agree in their basic assumptions.

Before we turn to the great achievements of quantum mechanics, especially in the form of quantum field theory, let us take a look at the invention of classical field theory, which can be viewed as another step for physics away from a direct contact with reality.

6

Fields

Of all of Newton's revolutionary physical ideas the most counter-intuitive and abstract had been that of the universal force of gravity. Acting as an attraction between objects distant from one another and without any physical agent transmitting it, such a force contradicted everyone's notion of what reality was like. It was regarded as repugnant by all, including initially its inventor. However, as time went on, physicists and philosophers got used to it and accepted it because of its remarkable success. When combined with Newton's laws of motion, it accounted for Kepler's laws for the motions of the planets and it even successfully predicted the trajectories and the recurrence of comets. Unease with the notion of action at a distance nevertheless remained when, at the beginning of the 19th century, it became urgent to explain the electrical and magnetic forces that had been discovered since the time of Newton. After André Ampère and Hans Christian Oersted had discovered that currents could produce magnetism, Joseph Henry and Michael Faraday had found that magnets could induce electric currents and Faraday was looking for a way to account for the effects of magnetism and electricity on remote bodies without resorting to Newtonian action at a distance.

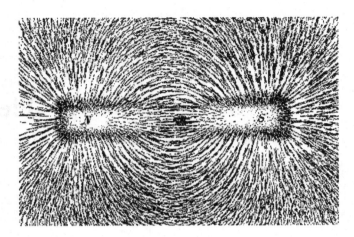

Figure 8.   Iron filings in the presence of a magnet.

The ingenious manner in which Faraday pictured the action of magnets and electric charges on one another at a distance was stimulated by the stream-like appearance of iron filings near a magnet (see Figure 8). He envisaged space being filled with rubber-band-like lines that transmitted the attractions and repulsions from one charge to another and from one magnetic pole to another. These lines of force were real for him, and he did not think of them as made up of particles. "If they exist, it is not by a succession of particles…," he wrote, "but by the condition of space free from such particles. A magnet placed in the middle of the best vacuum we can produce… acts as well upon a needle as if it were surrounded by air, water or glass; and therefore these lines exist in such a vacuum as well as where there is matter."[24] The resulting space-filling lines of force he called a *field*. The denser the lines of force, the stronger the field.

The advantage of this picture over the way Newton presented gravity was clear. Instead of having one charge directly exerting a force

---

[24] Quoted in Williams, L. Pearce, *Michael Faraday: A Biography*. New York: Da Capo Press, Inc., 1987, p. 450.

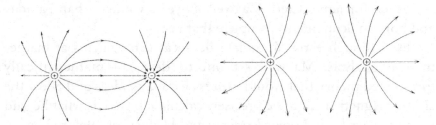

Figure 9. Lines of electric force, as Faraday imagined them, in the presence of two equal (right) and opposite (left) charges. (Reproduced with permission from J.T. Thomas, *Michael Faraday and the Royal Institution*, A. Hilger, 1991, p. 46.)

of attraction or repulsion on another at a distance the way Newton had one mass exert a gravitational attraction on another, Faraday imagined an electric charge producing an electric field everywhere in space and when another charge was placed at a distant point, it was the electric field at the very position of this charge that exerted a force on it. Similarly for magnetic forces: the poles of one magnet would produce a space-filling magnetic field that exerted a force upon a magnetic pole at the very point where this pole was located. What is more, since electric currents produced magnetism and magnets in turn affected currents, the electric and magnetic fields formed only separate aspects of a combined *electromagnetic* field.

Thus Faraday's notion of electric and magnetic fields was a totally imaginary mental construction whose sole purpose was to avoid the repugnant idea of action at a distance like Newtonian gravitation. In his mind's eye he saw the field as made up of lines of force like rubber bands, and he even speculated that all matter, rather than consisting of atoms, may be a particular manifestation of the space-filling lines of force. The question of whether the electromagnetic field itself corresponded to reality in any philosophical sense was of little concern to him, but his vision certainly had a great deal of explanatory power. However, Faraday's mathematical abilities were weaker than his imagination, and he was unable to carry his ingenious and enormously fruitful idea to a constructive conclusion.

It was left for James Clerk Maxwell, 40 years younger than Faraday and born in Scotland, to complete that task.

Even though early tutors had deprecated his talents at mathematical analysis, Maxwell set out to devise a mathematically formulated theory that would express Faraday's ideas and allow the electromagnetic field to be precisely calculated. Here is what he said in the preface to his book *Electricity and Magnetism*: "Before I began the study of electricity I resolved to read no mathematics on the subject until I had first read through Faraday's *Experimental Researches on Electricity*. I was aware that there was supposed to be a difference between Faraday's way of conceiving phenomena and that of the mathematicians... As I proceeded with the study of Faraday, I perceived that his method of conceiving the phenomena was also a mathematical one, though not exhibited in the conventional form of mathematical symbols... For instance, Faraday, in his mind's eye, saw lines of force traversing all space where mathematicians saw centres of force attracting at a distance; Faraday saw a medium where they saw nothing but distance; Faraday sought the seat of the phenomena in real actions going on in the medium, they were satisfied that they had found it in a power of action at a distance impressed on the electric fluids."[25]

In Maxwell's view the field would be expressed as a stress-like condition in a ubiquitous *ether*, and for this he employed a mechanical model of great complexity. All space would contain a vast number of very small contiguous cells that are free to rotate under the action of magnetic or electric force (see Figure 10) and act like vortices. The equations he set up for the electromagnetic field modeled in this manner were inspired by those for a fluid. "A conducting body may be compared to a porous membrane which opposes more or less resistance to the passage of a fluid; while a dielectric is like an

---

[25] Quoted in Campbell, Lewis and Garnett, William, *The Life of James Clerk Maxwell*. New York: Johnson Reprint Corporation, 1969, p. 516 (first published in London, 1882).

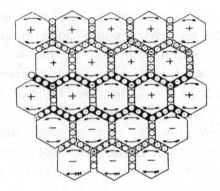

Figure 10. The cells making up the electromagnetic ether, as Maxwell imagined them.(Reproduced with permission from J. Hendry, *James Clerk Maxwell and the Theory of the Electromagnetic Field*, A. Hilger, 1986, p. 172.)

elastic membrane, which may be impervious to the fluid, but transmits the pressure on the one side to [the fluid] on the other."[26]

The result of Maxwell's detailed model-based reasoning was a set of complex partial differential equations known ever since simply as Maxwell's equations. In contrast to the way Newton mathematically formulated his force of gravity, directly giving the magnitude of that force from one body on another in terms of their mutual distance, the electromagnetic forces are characterized by means of differential equations, which, by means of differential calculus, state the manner in which the electromagnetic field at one point in space and time is influenced by the field at neighboring points. He thereby completely avoids any resort to action at a distance. Even light is an undulatory manifestation of this electromagnetic field, propagating from point to point at a fixed velocity, its sinusoidal waves of various wavelengths perceived by the human eye as different colors.

It took a number of years for physicists to be persuaded of Maxwell's theory. However, after Heinrich Hertz experimentally discovered the waves, longer than those of light, predicted by Maxwell's

[26] *Loc. cit.*, p. 541.

equations and now known as radio waves, his equations were generally accepted. The underlying complicated model of interacting vortices in the ether utilized by him for his theory, on the other hand, did not find much favor. But no matter; when Hertz wrote his book *Electric Waves,* he simply dismissed the model and announced "Maxwell's theory is Maxwell's system of equations"; and so it has been ever since. Even the very existence of the luminiferous ether, the stress of which Maxwell meant to describe by means of his equations, has since been discarded, made redundant by Einstein's Special Theory of Relativity.

The question now arises as to what extent the electromagnetic field described by Faraday and Maxwell can be regarded as "real." Even Einstein, much as he admired their theory and much as he insisted that physics should directly describe reality, recognized that their field theory represented "a step in the direction of constructive speculation which has increased the distance between the foundation of the theory and sense experiences."[27] The most persuasive argument for the reality of the electromagnetic field is the fact that Maxwell's equations imply that the field carries energy. As moving electrically charged particles produce an electromagnetic field and the same field accelerates charges immersed in it elsewhere, there is a continual exchange of energy between the field and the charges. If the field were ignored or regarded as fictitious, conservation of energy would be violated. Nevertheless, there can be no denying that field theory has "increased the distance between the foundation of the theory and sense experiences."

Abstract though Faraday's idea of the field may have been, and far from sense experiences, it turned out to be enormously fertile. Generalized for the theoretical description of later discovered forces other than electromagnetism, it became indispensable in physics, especially when combined with quantum mechanics in the form of quantum field theory. The kind of theory that Einstein spent the last

---

[27] Einstein, *Ideas and Opinions,* p. 305.

25 years of his life seeking to replace quantum mechanics was a field theory too. Even though he recognized the distance of its foundation from sense experience, this distance from reality was not nearly as great as that of quantum mechanics. However, he failed to bridge that distance and physics passed him by.

We now turn to the great achievements of quantum mechanics when its procedures were applied to field theory. It was quantum field theory that managed to do what classical physics never could, namely to account for, rather than merely accept and describe, the particulate structure of the world, a structure that turned out to be vastly more complicated than originally imagined. And it is important to recall that in contrast to its description of dynamics, which was probabilistic and remote from direct contact with reality, quantum mechanics, from its very beginning, and therefore quantum field theory as well, dealt with structural matters quite directly. There had been nothing probabilistic about the energy levels of Bohr's atom. Probabilities and their distance from reality itself entered the picture only when changes took place and radiation was emitted.

7

New Particles and their Quantum Origins

Quantum mechanics is a framework theory in the same sense in which Newton's laws of motion constitute a framework theory. Both need to be supplemented by specific laws governing the forces, or as we now say, the "interactions," between entities. In order for Newton to explain how the planets moved, he needed not just his equations of motion but also his law of universal gravitation governing the force between the sun and the planets. Similarly, in order for Schrödinger's equation to describe the functioning of the atom, the electrostatic forces among the electrons and between electrons and the nucleus have to be added to the quantum framework. Once that is done, the possible energies of the atom can be determined. Except for the enormous difference in scale, the forces in an atom very much resemble those in a planetary system, the nucleus taking the place of the sun and the electrons those of the planets. Even the way the electrostatic force decreases as a function of the distance is the same as that of gravity. But the quantum-mechanical result differs greatly from the Newtonian.

One of the fundamental differences between the results of a calculation of the possible energies of an atom via Newton's equations and via quantum mechanics is that, in addition to a continuous range of energies at which some of the electrons roam around freely, dissociated from the atom (the atom is ionized[28]), the latter allows only certain discrete energies. This is just what Bohr had postulated, and the energies he had specified were approximately correct.

Obtaining specific discrete energies, however, accounts for only half of Bohr's model. To complete the picture, the fact that the electrons are electrically charged has to be considered in the sense that they are therefore coupled to the ever-present ambient electromagnetic field. This complicates the calculations, but the effect is that the atom is completely stable only in its configuration of lowest energy, its "ground state"; in that state it can remain forever unless disturbed. In any other state it has a certain probability per unit time to descend from such an "excited" state to another state of lower energy, emitting the energy difference E in the form of radiation of frequency f such that $E = hf$, just as the second half of Bohr's model had assumed. Furthermore, quantum mechanics predicts the probability per unit time for this jump to occur, and therefore how long, on average, the atom would remain in its excited state. (Einstein had already provided a formula for this probability before quantum mechanics validated it.)

The most important feature added by quantum mechanics to the theoretical understanding of a system such as an atom is the absolute stability of the system in a precisely specified ground state. Furthermore, according to quantum mechanics, any helium atom in its ground state is fundamentally indistinguishable from any other helium atom in its ground state. Where classical physics allowed certain small imperfections and identifying marks, quantum mechanics permits none. An atom is completely specified by its "quantum numbers," including the energy level it is in: a helium atom is a helium atom is a helium atom. Note that here the otherwise

---

[28] An ion is an electrically charged atom, i.e., an atom with one or more electrons either missing or added.

probabilistic nature of the quantum theory is absent, nor is there any noticeable remnant of its distance from reality.

Quantum Field Theory

Let us now look at quantum field theory. Classical field theory had been the invention of Michael Faraday in the first half of the 19th century (see Chapter 6). Rather than having one electric charge directly exert a force on another at a distance, the way Newton had the sun attract the earth, Faraday assumed that the first charge produced everywhere a condition of space called an electric "field," and this field exerted a force on the other charge right there, where it was situated; similarly for a magnetic field. James Clerk Maxwell brought this novel notion to mathematical perfection. His equations completely defined the electromagnetic field — electricity and magnetism combined into one — produced by any system of electric charges and magnets, as well as the manner in which this field changed with time, moved in space, and acted on other charges. One of its oscillatory manifestations was light; others were, as soon thereafter verified by Heinrich Hertz, what we now call radio waves and microwaves, of much lower frequency than light, as well as X-rays and gamma rays, both found later and of much higher frequency.

In order to understand the existence of photons it was necessary to "quantize" Maxwell's equations, just as the quantizing of Newton's equations was needed to understand the energy levels of atoms. This process begins with the same analysis of Maxwell's equations that led to solutions describing light of a specific frequency.[29] Solutions were sought that oscillate with a fixed frequency, like the motion of a simple pendulum, as though the electromagnetic field were made up of an infinite collection of "harmonic oscillators" (the physicists' name for simple pendulums).

[29] This is called Fourier analysis after the French mathematician Joseph Fourier.

To quantize the equations amounts to treating each of these mathematical pendulums of a given frequency of oscillation by the procedures of quantum mechanics, and the result is that the allowed "spectrum of energies" is not continuous but discrete, just like the levels in Bohr's atom. The spectrum of a harmonic oscillator, however, has a very unusual structure: it consists of infinitely many levels, all equidistant. In other words, the energy difference between the first and the second level is the same as that from the fifth to the sixth, as from the 10th to the 11th, etc. And this energy difference E is related to the frequency f of the pendulum's oscillation by Planck's formula $E = hf$. One may therefore think of the oscillator in its first excited state (the first state higher than the ground state) as containing one quantum of energy hf; when it is in its fifth excited state as containing five quanta of energy hf; when in its 20th excited state as containing 20 quanta of energy hf, etc. The energy of the electromagnetic quantum field oscillating with the frequency f is thus made up of quanta, each with the same energy hf, and they are moving with the speed of light. These are Einstein's light quanta, or photons. They originate from the quantum-mechanical treatment of Maxwell's equations just the way oscillating light waves originate from the classical treatment of the same equations.

Analogous to the way Maxwell's field theory, when quantized, led to photons, so other quantum field theories led to a number of other particles. In addition to predicting their existence, quantum field theory explained *why* the world consisted of particles, and why these particles had the properties they were found to have.

The First New Particles

The first prediction of the existence of a hitherto unknown particle was based on a need to make quantum mechanics conform to Einstein's Special Theory of Relativity. The theory originally set up by Heisenberg, Schrödinger, Dirac, and Born had not been consonant with the demands of relativity, but for its first applications this had been unnecessary. There were, however, some occasions in atoms on which electrons moved — when envisioned in a classical manner — almost with the speed of light, and there would soon be more such occasions in experimental devices. This meant it was imperative for relativity[30] to be taken into account. To perform this task, Dirac therefore invented a new quantum mechanical equation for electrons, and it did its job both beautifully (difficult though it was to define exactly what that meant, beauty was an important criterion for Dirac) and with great experimental success. When he subjected this equation to quantization analogous to what had been done to Maxwell's — in this case it was called second quantization because his original equation was already a quantum-mechanical wave equation analogous to Schrödinger's — the electron emerged as a particle with all the right properties, just as the photon had emerged from Maxwell's equations. What is more, he found to his surprise that his equation had not only solutions describing electrons, but also solutions describing particles just like electrons, with the same mass but opposite electric charge: they were *positive*. No one had ever seen such a particle. However, it took only a few years for it to be found by the American physicist Carl Anderson in 1932 among cosmic rays bombarding the earth from outer space. Its track was visible in a cloud chamber, the device C.T.R. Wilson had invented 20 years earlier, looking exactly like that of an electron, but in a magnetic field it curved in the opposite direction. Named *positron*, it was the first of what would later be called *antiparticles*.

Just before Anderson's discovery, two other new particles had been found, one unexpectedly experimentally discovered and the existence of the other theoretically predicted, though not like that of the positron.

The neutron had been conjectured by Rutherford, for it was needed to make more sense out of the atomic nucleus he had

---

[30] This must not to be confused with conforming to the General Theory of Relativity!

discovered, his "fly in the cathedral." The numerical position of an element in the periodic table was understood to correspond to the number of electrons in its atoms, and these had to be equal to the number of protons in the atoms' nuclei so as to make the whole atom electrically neutral. But the atomic weights of all elements heavier than hydrogen were usually about twice their atomic positional number. The nucleus had to contain other particles, just as heavy as protons, without adding to the charge. One notion was to assume that the nucleus also contained electrons, as many as there were extra protons, so as to neutralize them. That would also conveniently account for the beta rays emitted by radioactive elements, which surely had to come from the nucleus. But a simpler idea was to imagine that the nucleus contained neutral particles, each of the same weight as protons. Rutherford was confident such a neutral particle would eventually be found. And it was in 1932, by the British physicist James Chadwick.

Particles that are electrically neutral are much harder to detect than charged ones because the signals that give away the presence of a tiny particle — much too small to be visible under a microscope — are always caused by its electric charge. A neutral particle leaves no trace in a cloud chamber, it leaves no image on a photographic plate, and it exerts no electrostatic forces. The presence of neutral particles can only be inferred from the absence of evidence for a charged particle needed for conservation of energy and momentum. So Chadwick directed a stream of alpha particles at a target made of beryllium. In contrast to an earlier experiment in which Rutherford had shot alpha particles at nitrogen gas and found the emission of protons, beryllium was known from previous experiments to react to such bombardment by emitting some kind of penetrating neutral rays, assumed at the time to be gamma rays. Allowing these rays to strike a cushion made of paraffin, a wax made of hydrocarbons, he had a final detector identify the particles released by these secondary collisions as ionized hydrogen atoms; moreover, the detector could even estimate the speed of these emitted ions. The rays emitted by the beryllium, he concluded, had to consist of particles just as heavy as the fast-moving protons they knocked out of the paraffin; gamma rays would not be able to accomplish that. The results of his experiment allowed him even to measure the mass of this neutral particle. He had finally found the neutron Rutherford had expected. Its mass, as determined by Chadwick, was just a smidgen greater than that of the proton. (The tiny amount, less than 1%, by which its mass exceeded that of the proton would later turn out to be crucial in explaining radioactivity.)

The year before Chadwick's discovery of the neutron, another neutral particle had been introduced into physics, although the experimental confirmation of its real existence would take many years. When a radioactive element such as uranium 238 emits nothing but its daughter element thorium 234 and alpha rays, detailed experiments showed that all the emitted alpha particles had the same energy. This was exactly as it should be. The energies of the emitted thorium and helium atoms can be very simply calculated by the laws of conservation of energy and momentum once the masses of the three atoms are given. However, in the radioactivity of those elements that seemed to emit only beta rays in addition to their daughter elements, the emitted electrons were found to have a whole range of possible energies. This was so puzzling that Bohr was ready for another revolution by questioning the law of conservation of energy. (There seemed to be no change in mass involved that might account for that.) The incipient revolution was nipped in the bud by Wolfgang Pauli with the suggestion that in the decay there must be another particle given off that remained undetected because it was neutral and very light (as light as or lighter than the electron). If that were the case, the energy of the emerging electrons would not be unique but depend on the momentum and the energy of the undetected fugitive particle. Enrico Fermi called it the *neutrino*: the little neutral one in Italian. (In our present terminology this particle is actually an antineutrino.) Even though nobody seriously doubted its existence after many different beta-decay observations, because the neutrino carries no electric charge and interacts only very feebly with matter, its presence was extremely difficult to verify directly. It took about 20 years before Fred Reines and Clyde Cowan finally did so.

All these particles have certain additional properties of great physical significance. In the first place, they have an intrinsic angular momentum called spin, as though spinning around a built-in axis. According to quantum mechanics, the magnitude of such spin can only be an integral (whole-number) multiple of Planck's constant h, or else a half-integral multiple of h. The spin of the electron, as well as that of the positron, is $1/2h$ (see Chapter 3 for the experimental discovery of this fact by Stern and Gerlach), as are the spins of the neutron and the neutrino, and that of the photon is h. Furthermore, electrons (as well as positrons) are subject to Pauli's "exclusion principle," whereas photons are not. The exclusion principle prohibits any two of these particles from being in the same state. This is the reason why the electrons in a heavy atom have to occupy higher and higher energy levels rather than all congregating on the lowest level, which explains much of the predictive power of the periodic table of the elements. The special statistical properties of assemblages of indistinguishable particles subject to the exclusion principle were first elucidated by Paul Dirac and Enrico Fermi; these statistics are therefore called Fermi–Dirac statistics and the particles are called fermions. The statistical properties of fundamentally indistinguishable particles like photons, not subject to the Pauli principle, on the other hand, were spelled out by Einstein and the Indian physicist Satyendranath Bose (see Chapter 3); these are named Bose–Einstein statistics and the particles are called bosons.

We already saw in Chapter 3 why indistinguishable particles are subject to different statistics than distinguishable ones. If two particles A and B are to be randomly distributed among three boxes, we saw that there are nine ways if they are distinguishable, but only six if they are indistinguishable. If, furthermore, the particles have to obey the exclusion principle and no two can be in the same box, there are only three. This is why there are ordinary, so-called Maxwell–Boltzmann statistics for distinguishable particles (those developed for classical statistical mechanics), and Bose–Einstein and Fermi–Dirac statistics for indistinguishable ones.

Electrons (and positrons) are fermions and their spin is $1/2h$, while photons are bosons and have spin h. This is no accident; it is one of the fundamental achievements of quantum field theory to imply a universal connection between spin and statistics: particles of half-integral spin are necessarily fermions, while those of integral spin are bosons. This rule applies even to systems made up of several particles: composites such as nuclei and atoms made up of even numbers of fermions (whose spins therefore add up to whole-number multiples of h) also act like bosons, a fact that has important consequences, as we shall see later. It will turn out that the particles that form the fundamental building blocks of matter are all fermions; the fundamental bosons are the quanta of what might be called force fields, just as photons are the quanta of the electromagnetic force between charged particles.

What Holds the Nucleus Together?

The great problem that preoccupied physicists in the late 1930s was the atomic nucleus. What accounted for the disintegration of the nucleus in radioactivity? And the most pressing question: what force held it together? After all, the protons in the nucleus were all positively charged and therefore repelled one another, and this repulsion increased at short distances. There had to be a strongly attractive force among protons, able to overcome the electrostatic repulsion among them, but also acting between protons and neutrons as well as among neutrons, so as to explain the cohesion of the atomic nucleus. Furthermore, this force had to be of short range, reaching no farther than the size of the nucleus, as no evidence of it had ever been detected outside.

The first theorist to propose, in 1934, a field theory that would lead to such a force was the Japanese physicist Hideki Yukawa. The model he followed was quantum electrodynamics, usually referred to as QED, the combination of the quantized Maxwell equations and the second-quantized version of Dirac's equation that physicists at the time were struggling to make into a consistent theory. (They did not succeed until more than 15 years later, with brilliant results experimentally confirmed with unprecedented accuracy.) However, by the rules of quantum field theory, the place of the massless photon, the quantum of the long-ranged electrostatic force, had to be taken by a massive particle so as to account for the short reach of the strong nuclear force. Indeed, from the range of that force, roughly equal to the size of the nucleus, the mass of Yukawa's "U-particle" (later called *meson*) could be estimated to be several hundred times that of the electron. The search was on to find evidence for its existence.

This was a time when many teams of physicists used cloud chambers to make detailed observations on cosmic rays, streams of particles impinging with very high speeds on our atmosphere from outer space. And several of these teams found particles of roughly the right mass. However, after several years of uncertainty, it became clear that the "mesotrons" seen in cosmic rays could not be Yukawa's mesons. They were found to be able to penetrate through deep layers of earth, indicating that they did not interact strongly with atomic nuclei as the mesons had to if they were to explain the nuclear force. These unanticipated particles — "who ordered these?" was the reaction of the prominent physicist Isadore Rabi to their discovery — were later renamed *muons*, their properties much like those of electrons but about two hundred times heavier. It took another 10 years for the real Yukawa mesons to be found by two teams of physicists in Bristol and London. Renamed *pions*, they turned out to exist in three varieties, positive, negative, and neutral, somewhat heavier than muons, and endowed with spin zero. Physicists spent the next several decades experimentally studying the energy levels and other properties of nuclei, and attempting to understand them on the basis of Yukawa's force transmitted by these pions, generally with success.

The New Accelerators

In order to study the properties of the atomic nucleus and to search for the existence of more particles, it was no longer sufficient to rely on streams of cosmic rays coming in from outer space or on the emissions of radioactivity. More control was required and particles were needed in larger numbers. Since the method of accelerating them always relied upon the force exerted on an electric charge by an electric field generated by a voltage difference, new instruments were invented to produce higher and higher voltages. So prevalent was this procedure of speeding up particles that it became customary to use the voltage required to push an electron to a given energy as a unit of energy: 1 eV (electron volt) is the energy of an electron accelerated by the electric field generated by a potential difference of one volt; 1 MeV is the energy of an electron accelerated by the electric field generated by a potential difference of one million volts; eventually it would be convenient even to use the unit GeV, one billion electron volts.

There were two reasons why particles of increasingly higher energies were needed for these investigations. The first was that in order for a heavy particle like the proton to penetrate the nucleus, it had to overcome the powerful repulsion exerted by the positively charged protons that resided there; electrons, which would not be subject to this handicap, were too light to knock anything out of the nest of protons and neutrons (the generic name *nucleon* would later be used for both). The second reason was that the most fertile method of finding new particles was to produce them in high-energy collisions. That it is actually possible to *produce* particles is the direct implication of Einstein's equation, $E = mc^2$, the inter-convertibility of mass and energy. In a sense, mass is just one form of energy.

Although he did not state it explicitly in his paper, nor is there any reason to think Einstein realized this implication at the time, this meant that provided all applicable conservation laws were satisfied, particles could be produced "out of nothing but energy."

To avoid misunderstanding, let us be clear about terminology. Even though photons may be regarded as particles, they are not *matter*. All matter is characterized by having a rest mass, that is, when brought to rest it has a certain mass m — this is what we call its mass in ordinary language — and its energy, according to Einstein's equation, is mc^2. A photon, on the other hand, can never be found still; in every reference frame it moves with the speed of light. Even though the energy E of a certain amount of electromagnetic radiation is equivalent to a mass of $m = E/c^2$, this radiation is distinct from matter. It is, in a sense, pure energy, which nevertheless can be converted into matter, at which point some — or all — of it attains rest mass. Einstein's equation should therefore not be interpreted as obliterating the distinction between radiation and matter; it merely implies that radiation and matter are inter-convertible.

The possibility, implied by Einstein's theory of relativity, of actually producing particles out of "nothing but energy" was first exemplified in quantum electrodynamics. QED can be used to calculate, in certain specified circumstances, the probability for electromagnetic radiation (a photon) to produce an electron–positron pair (conservation of charge requires that it be such a pair of positive and negative charges of equal magnitude), *provided* the photon has enough energy. The minimum energy needed would be just enough to produce both members of the pair at rest, without any additional kinetic energy; this would require the energy $2mc^2$, where m is the mass of the electron as well as of the positron. In the new energy units, including the factor c^2, the mass of the electron is about 0.5 MeV/c^2; so the minimum energy required to produce an electron–positron pair is about 1 MeV. To produce particles as heavy as protons, energies in the GeV range are necessary.

The first machines to produce the needed high voltages were invented by the Englishman John Cockcroft and the Irishman Ernest Walton, both working at Rutherford's Cavendish Laboratory. Their electrostatic generator managed to make a voltage of 710,000 volts, enough for them to produce helium by bombarding lithium with accelerated protons: they "split the atom" and produced sensational news. The American physicist Robert van de Graaff invented another kind of electrostatic generator that was able to produce 5 million volts. These bulky machines, however, were soon superseded by the invention of the cyclotron.

The American physicist Ernest Lawrence had the brilliant idea of accelerating charged particles by letting them run in circles and giving them a small kick every time they came around. Its path bent by a strong magnetic field, a positively charged particle would start moving slowly in a small circle inside two pita-bread shaped metal pockets separated by a gap, one charged positively, the other negatively (see Figure 11). As it moves through the gap from the positive

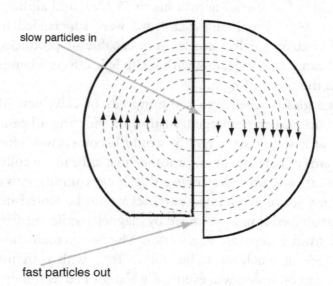

Figure 11. Sketch of the top view of a cyclotron.

shell to the negative, the electrostatic force accelerates it and increases the radius of its orbit; by the time the particle reaches the gap again halfway around, the voltages on the shells have been reversed, the previously negative shell has become positive, and the particle gets another kick. This process is repeated many times until its orbit reaches the outer perimeter and the particle has attained a very high speed. What makes this device possible is that, according to Maxwell's equations, as the particle accelerates along its spiral-shaped path, the time of each circuit remains the same. Therefore the frequency with which the voltages have to be reversed stays constant during the entire process of acceleration.

The first cyclotron Lawrence built for himself was a small table-top device, 4.5 inches in diameter, with 1,800 volts between the two "D"s, which would accelerate protons to an energy of 80,000 eV. By 1931 he had built an 11-inch cyclotron that achieved 1 MeV. Ultimately a cyclotron would be built as large as 88 inches in diameter, capable of accelerating protons to 55 MeV and alpha particles up to 140 MeV. Eventually cyclotrons were superseded by other kinds of circular particle accelerators capable of producing more focused beams and of dealing with relativistic effects when particles approach the speed of light.

As energies reached into the hundreds of GeV, new ideas for speeding up particles emerged. Instead of shooting a beam of fast particles at a stationary target, it would become more effective to shoot beams of particles at one another, just as head-on collisions of speeding cars are more destructive than a car running into another at rest. One set of accelerated particles would be stored in a large doughnut-shaped ring surrounded by magnets, while the other beam emerged from a separate accelerator. The Superconducting Super Collider (SSC), which was to be built in Texas with a circumference of more than 50 miles, was eventually abandoned as too expensive.

The circular tunnel for the Large Hadron Collider (LHC) accelerator at the CERN laboratory in Geneva, Switzerland, whose construction was finished in 2008 is about 17 miles long. The search for understanding the properties of submicroscopic particles and for finding new ones has transformed physics into Big Science.

8

Atoms, Inside and Out

The drive to understand the quantum-mechanical properties of the atomic nucleus had intensified during the 1930s, and some of the new knowledge would turn out to have momentous implications. In 1938, the two radiochemists Otto Hahn and Fritz Strassmann in Germany performed an experiment hoping to produce "transuranic" elements, i.e., elements heavier than uranium. The result left them mystified. Irradiating uranium with neutrons (neutron rays could be produced by bombarding beryllium with alpha rays, as Chadwick had done, as well as by other reactions of a similar kind), they found that they had produced barium, a much lighter element. Lise Meitner, a physicist who had been a member of their team for many years, was not available for advice because, as a Jew, she had recently been forced to leave Germany and was now in Sweden. When Hahn contacted her for help in understanding how they could possibly have produced a light element like barium, she and her nephew Otto Frisch, a physicist living in Britain, concluded that there was only one conceivable explanation: the neutrons impinging on the uranium nuclei had caused the latter to split, with barium one of the fragments, a process that came to be called nuclear fission.

Rumors of this discovery sent a tremor of anxiety through the international physics community, because all knowledgeable physicists realized its disturbing implications: the splitting of a uranium nucleus by a neutron would produce as byproducts more neutrons, which in turn would cause a larger number of surrounding uranium nuclei to fission, producing a snow-balling cascade that could lead to an enormous explosion. The political atmosphere in 1938 in Europe was so tense that to have Hitler's Germany in possession of a military weapon of such magnitude was a frightening thought. Fortunately, as the contingencies of history unfolded, it was not Germany but the United States that was able to develop the "atomic bomb."

Where does the enormous amount of energy come from? The mass of the uranium nucleus, made unstable after absorbing a slow neutron — Enrico Fermi, performing experiments in Rome, Italy, had discovered that slow neutrons are more efficient at causing nuclear transformations than fast ones — was somewhat larger than the sum of the masses of the fragments into which it splits. In the process of fission this excess mass is converted into kinetic energy of the fragments and gamma rays — both $E = mc^2$ and quantum mechanics at work. Even though the mass excess m for each fission event is small and so is the corresponding energy, the number of atoms in a small chunk of uranium is very large thus producing a huge amount of energy. Nuclear fission could be used either in uncontrolled form for explosive purposes or, when controlled by arranging to have most of the excess neutrons absorbed by some damping material, for the production of useful energy. It is a clear demonstration of the practical importance of the atomic constitution of matter and of the quantum-mechanical conversion of one form of energy into another. In 1942, Fermi at the University of Chicago — he had recently fled Italy — for the first time constructed a nuclear reactor for the production of energy by controlled nuclear fission.

Fusion was another nuclear process capable of generating energy. In this case the nuclei of two light elements fuse to form a heavier one whose nucleus is somewhat lighter than the sum of the weights of the initial ones, and the loss in mass is made up in part by the ejection of particles such as protons and neutrons and in part by conversion into kinetic energy of the ejected particles and gamma rays, again by Einstein's equation. This process, however, requires a large amount of kinetic energy to begin with so that the two initiating colliding atoms can overcome their electrostatic repulsion for the nuclei to come together. Promising though it is for useful energy production, this technique has not yet been tamed sufficiently to fulfill its promise. On the other hand, in an extremely hot environment many atoms move fast enough for fusion to work; under these conditions the result is called a thermonuclear reaction.

The Energy Source of the Stars

The interiors of the sun or other stars, where the temperature is millions of degrees, seemed to be just the right place for such reactions to take place; so perhaps thermonuclear reactions were the source of the energy that kept the stars hot. Hans Bethe proposed two specific sequences of reactions and calculated — employing, of course, quantum mechanics — the probabilities for them to occur in the solar environment. One, called the pp cycle, begins with the fusion of two protons, i.e., hydrogen nuclei, and, after three steps, ends with an alpha particle, i.e., a helium nucleus, and two protons, which are then available to repeat the same process. The other is more complicated but also begins with the fusion of hydrogen and ends with the production of helium. These thermonuclear reactions, he theorized, can produce enough energy to keep the sun and other stars shining for many billions of years.

How do we know that Bethe's processes are in fact the source of the light emission of the sun? Both processes produce as byproducts neutrinos of quite specific energy distributions, and these should be able to escape from deep inside the sun to be observed here on earth. After some years of confusion and uncertainty, they were found — recall that neutrinos are very difficult to detect because they are neutral and interact very weakly with anything — just as many as would correspond to what was known about the interior of the sun. We therefore have direct observational evidence of the correctness of Bethe's theory.

The idea of thermonuclear reactions in the stars led to other fruitful theories. The Indian-born astrophysicist Subrahmanyan Chandrasekhar used the consequences of Bethe's theory to explain the history of the development of stars and why some of them give rise to the spectacular explosions we see as supernovas.

Explaining the Origin of the Elements

The next several decades were spent by many physicists trying to explain the structure of atomic nuclei and the details of the force transmitted by Yukawa's mesons. The problem was much more complicated than the corresponding one for the atom because the force holding the nuclear constituents together was much stronger and because, in contrast to the atom, there was no heavy attractive center with much lighter satellites circling about; here all the particles, the neutrons and protons, attracting one another, were almost equally massive. Nevertheless, the quantum physics of the principal nuclear phenomena — such as their energy levels and gamma-ray emission during transitions, nuclear transformations initiated by collisions with alpha particles, nuclear fission, and nuclear fusion — became in principle well understood. Radioactivity was another story, which took longer to unravel, as we shall see.

Understanding nuclear physics turned out to be crucial for a rational explanation of the development of the world. Nuclear reactions not only helped us understand the energy source of the sun and the stars as well as their evolution, but they also served to explain the origin and abundances of the elements in the universe. If the young universe contained nothing but electromagnetic radiation and hydrogen nuclei, i.e., protons, the question was, how did all the heavier elements come into being, and why did the stars contain these in the relative abundances in which we find them now? (The numerical evidence for these abundances was based on the intensities of the various spectral lines characteristic of the atoms of each element observed by astronomers. These intensities were obviously proportional to the number of atoms of a given element emitting the radiation.)

The answers to these questions rested in part on a discovery about the neutron that would also turn out to be basic to the understanding of radioactivity. Surprisingly enough, the neutron itself was found to be subject to beta decay. With a half-life of about 10 minutes, it decays into a proton (which, as you may recall, is slightly lighter than the neutron), plus an electron, and an anti-neutrino. The reason why this instability of the neutron does not cause all nuclei containing them to be radioactive and to decay with a half-life of 10 minutes is that Pauli's exclusion principle prevents it: the proton emitted in the decay has very little energy (the mass difference is very small), and all the needed nuclear energy levels for it to move to are usually occupied. The variety of half-lives of radioactive elements is caused by intra-nuclear processes responsible for energy levels available for the newly created proton, but the original source of beta radioactivity is always the decay of a neutron.

Returning now to the creation of the elements, beginning with nothing but hydrogen, the Russian-born George Gamow suggested that it might all happen by means of two kinds of processes. If the nucleus of an element in the nth position of the periodic table, i.e., one with n protons, captures a neutron that then decays into a proton, while the electron and the neutrino escape, the new nucleus has $n + 1$ protons and therefore belongs to an element in the $(n + 1)$st position. Together with fusion reactions, this might be able to ratchet up the ladder of elements all the way to uranium. William Fowler and Fred Hoyle followed this suggestion and performed detailed quantum-mechanical calculations of the probabilities of these processes actually occuring — a most arduous procedure — and were able to account quite accurately for the relative distribution of all the elements in the world as observed by astronomers. A remarkable astronomical verification of our understanding of nuclear processes! At the same time, nuclear physics had turned out to be an indispensable tool for learning the evolution of the universe.

Radioactivity had other far-reaching implications. For one thing, the fact that each radioactive element has a characteristic half-life would lend itself to a most useful method for dating the origin of objects such as rocks or ancient artifacts. For another, it became clear that since radioactive elements were distributed throughout the interior of the earth, the energy they generated in the course of their decay should produce heat and thus contribute to keeping the earth warm. This fact should be taken into account when estimating the age of the earth, but it had been unknown during the 19th century. One of the most powerful arguments against Darwinian evolution at that time had been calculations by the two prominent scientists Kelvin and Helmholtz, based on the earth's rate of cooling since its birth, that the earth could not be older than a few hundred million years (in Helmholtz's calculation, no more than 25 million). This was much too short a time for the evolution of all the species to have taken place. The discovery of radioactivity, and the realization that it helped to keep us warm, changed the estimate to about 1.6 billion years, more than enough of the required time for biological evolution. The implications of what was going on inside Rutherford's fly in the cathedral were astonishingly wide.

Explaining Radioactivity

Now what about the origin of radioactivity? Understanding alpha decay was a relatively straightforward matter. The nuclear force happens to be such that the four constituents of an alpha particle, two protons and two neutrons, are particularly strongly bound together, even when they find themselves part of a larger nucleus. What prevents the alpha particle from escaping the larger nucleus, of course, is the strongly attractive Yukawa force, the short range of which creates an insurmountable barrier of a certain width — insurmountable, that is, according to classical Newtonian physics. Quantum mechanics, on the other hand, as the three physicists

George Gamow, Ronald Gurney, and Edward Condon proposed, only says that penetrating this wall is unlikely; it allows us to calculate the probability for "tunneling through" such a barrier, and thereby to calculate the average length of time before such an alpha decay would take place. Thus alpha radioactivity could be explained and the half-lives theoretically determined.

The beta radioactivity of the neutron was another matter entirely. Fermi was the first to construct a promising theory, though it eventually had to be abandoned. The difficulty originated in the fact that in the decay of the neutron three particles were involved, a nucleon, an electron, and a neutrino, rather than two, as in the interaction of the electron with a photon. As a result it turned out to be difficult to construct a consistent quantum field theory roughly modeled on QED, in which calculations could be done without the appearance of infinities: the three-particle theory could not be "renormalized." It took another 30 years for a renormalizable theory to be developed.

In the 1960s, Sheldon Glashow, Abdus Salam, and Steven Weinberg independently managed to unify the "weak interaction" responsible for beta decay with the electromagnetic one into a universal "electroweak" force transmitted by the photon and three massive bosons, one neutral, called Z^0, and two charged, W^+ and W^-, analogous to QED. (They thereby avoided relying on a three-particle theory and its associated problems.) When a sufficiently powerful accelerator had been constructed at the CERN laboratory in Geneva, Switzerland, all three newly proposed particles were eventually experimentally found. They turned out to be quite massive, the two Ws each having a mass of about 80 GeV/c^2, and the Z of about 91 GeV/c^2. (Remember, the mass of a proton is about 1 GeV/c^2.) All three are unstable, with a half-life as short as about 3×10^{-23} seconds.

The assumed universality of the electroweak theory, however, gave rise to a new problem: it led to the quantum-field theoretical prediction that the muon, known to be unstable, should decay into an electron while emitting a gamma ray photon. Such a decay, long sought by experimenters, had never been seen. Its absence could be explained only if the neutrino that interacted with the muon was different from the familiar one that interacted with the electron in beta decay. And indeed, this "two neutrino hypothesis" was confirmed in 1962 both at Brookhaven National Laboratory and at CERN in Geneva. The number of different particles found in nature was beginning to proliferate. But before relating the discoveries of more and more new particles, we need to pause and look at the consequences of quantum mechanics for the properties of solid matter.

Properties of Solid Matter

After the acceptance of atoms as the fundamental constituents of matter it was, of course, clear that all the properties of solids had to be explained in terms of the properties of atoms. Most solid materials came in the form of crystals, and the various shapes of these had been studied by mineralogists for a long time, but beginning in the 20th century, the structure of these crystals was investigated in more detail by means of the newly discovered X-rays, and later by means of electron and neutron beams. The X-ray diffraction[31] pictures showed that the atoms in a crystalline solid were arranged in a very orderly fashion, essentially at the corners of Platonic solids. This explained the specific, well-defined angles of the planar faces on the surfaces of crystals, and quantum mechanics was able to account for the atomic arrangements. What remained was to explain phenomena like magnetism and electrical conductivity.

A solid is called paramagnetic if, when placed in a magnetic field, it becomes magnetized in the same direction as the field; diamagnetic

---

[31] Diffraction is the phenomenon of alternating constructive and destructive interference of waves that caused the fringes in the experiment of Thomas Young. That electrons and neutrons suffer diffraction too is explained by quantum mechanics.

if magnetized in the opposite direction. Recall that in the 19th century Michael Faraday had discovered the connection between magnetism and electricity and Maxwell had fleshed out this connection explicitly by means of his equations. These equations made reference to moving charges but not to magnetic poles.[32] It therefore was initially assumed that the magnetism of solid materials, made up as they are of atoms or molecules, which in turn contained electrons, was caused by the motion of electric charges. This was correct, as experiments revealed, for diamagnetic materials, but not for paramagnetic ones.

Electrons, you may recall, have an angular momentum called spin, as though they are rotating about an axis through their center. As a result of this spin and the fact that they are electrically charged, they are in effect little permanent magnets of a specific strength called a "Bohr magneton," which point in the direction of the axis of their spin. In an atom, Pauli's exclusion principle allows two electrons with opposite directions of spin to occupy each energy level. The magnetic fields produced by these two electron-magnets pointing in opposite directions therefore cancel each other out, except when the last level is occupied by only one electron. As a result, atoms with an odd number of electrons are little magnets of their own, with the same strength as that of the last electron. This accounts for the paramagnetism of such materials.

More familiar than either paramagnetism or diamagnetism, ferromagnetism is the much stronger magnetic property of the elements iron, cobalt, and nickel that turns them into permanent magnets after being placed once in a magnetic field. These materials are made up of microscopic crystalline domains, each containing

---

[32] Magnets, which Maxwell's equations accommodates, always have two poles of opposite polarity; single, isolated poles do not occur in nature and Maxwell's theory has no provision for them. In that sense, magnetism is treated differently from electricity in Maxwell's theory.

many atomic magnets. When placed in a magnetic field, all these atomic magnets line up along the direction of the field, and when the field is turned off they remain that way, forming a permanent magnet. As the temperature is increased, the heat motion of the magnetic domains becomes more and more violent, their directions more and more shaken up, and the material loses its magnetism. However, Pierre Curie, the husband of Marie, discovered a strange phenomenon: ferromagnetism suddenly disappears at a fixed critical temperature now named the Curie point. This is what physicists call a phase transition. Why that should be so was very difficult to explain mathematically from thermodynamics, especially since no collection of a finite number of tiny magnets could possibly behave in such a discontinuous fashion. It would require an infinite number of them to produce any abrupt change at a fixed temperature, and of course the size of Avogadro's number was such that real materials might as well be assumed to contain an infinite number of atoms.

What physicists often do when nature confronts them with a very difficult mathematical problem is to replace the real problem with a simplified model that contains all the essentials of the real situation but that is easier to handle mathematically. In this instance the model was proposed by the German physicist Ernst Ising. It consisted of an arrangement of equally spaced little arrows, each of which could only point up or down and would exert a certain amount of force on its nearest neighbors to make them line up parallel to itself (see Figure 12). The entire "magnet" would then be set to vibrate as appropriate for a given temperature. The questions to be answered were whether there would be "long-range order," i.e., large domains would form in which all the arrows would be lined up parallel, and whether this long-range order would abruptly disappear at a certain fixed temperature.

Even this toy model presented a formidable mathematical problem. In one dimension, with all the arrows centered on and at right angles to a single, infinitely long straight line, it turned out to be easy

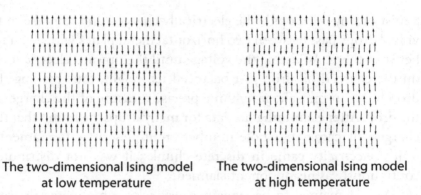

The two-dimensional Ising model at low temperature	The two-dimensional Ising model at high temperature

Figure 12.   The two-dimensional Ising model below and above the Curie point.

to see that there would be long-range order, but no phase transition, i.e., no Curie point. The much more difficult problem in two dimensions, that is, in an infinite plane, was finally solved, and the Norwegian-born chemist Lars Onsager was able to prove that there was indeed a definite temperature above which the long-range order established at low temperature disappeared. The problem in the real three-dimensional world has not been solved to this day, even for this vastly simplified model. In other words, it is still not mathematically understood why there is a Curie point at which a ferromagnetic phase transition takes place.

As far as the electrical conduction in metals and semiconductors is concerned, it had already been found experimentally before the discovery of electrons that the carriers of electricity had to be negatively charged particles. Once it was understood that atoms contained electrons, these were of course recognized as the carriers. In addition, there had been an important series of experiments carried out by the American physicist Robert Millikan during the years 1909 to 1913, which clearly established, against strong counter arguments from another physicist who thought he had proved the opposite, that electricity came in discrete negative units equal to the charge of the electron. The way Millikan had done this was to allow a mist of tiny oil drops to be electrically charged and to make individual drops pass between two horizontal metal plates with a voltage between them. Adjusting the voltage until the upward electric force on the charged oil drop just balanced its weight, thus making the drop hover in mid-air, gave him a precise measure of the charge on the drop. After examining his data for many drops, he found that the charges all came in whole-number multiples of a fundamental unit[33]: electricity came in discrete chunks; it was not distributed continuously as his opponent claimed.

What remained to be explained was why some materials were conductors and some semiconductors, and why it was that the conductivity of metals decreased with rising temperature while that of semiconductors increased, as had been observed in experiments. Efforts to understand these properties of materials by means of quantum mechanics gave rise to a new branch of physics called condensed-matter physics.

A metal, it became clear, consists of a fixed array of ionized atoms, their ejected outer electrons free to roam. These free electrons, subject to the Pauli exclusion principle and Fermi–Dirac statistics, move about as a "Fermi gas." At energies below the "Fermi surface" all energy levels are occupied, while above that surface, the electrons are able to move freely along "conduction bands," quantum mechanically accounted for by the perfect regularity of the crystal lattice. The electrical resistance of the metal is caused entirely by impurities, which destroy the pristine regularity of the crystal, introducing a bit of disorder that increases with rising temperature.

The properties of semiconductors such as germanium and silicon, situated between insulators and conductors, are quite different

---

[33] Millikan's treatment of his experimental data has been subject to criticism, though his result was correct. For details, see Gerald Holton, *Einstein, History and Other Passions*, Reading, MA: Addison-Wesley Publ. Co., 1996, p. 70. To this day, students in introductory physics classes are asked to repeat this experiment as an exercise, which they do with some difficulty.

from those of metallic conductors. (These materials are important for the construction of transistors, ubiquitous in practically all electronic devices.) Whereas a metal would be a perfect conductor if it were not for the impurities destroying the perfect regularity of its crystal lattice, a semiconductor owes its conductivity entirely to its impurities. In both kinds of materials, however, the important point is that the underlying explanation of their electrical conduction is based on the detailed properties of their atomic constituents.

Superconductivity

There is another quite remarkable kind of electrical conduction, which was discovered by the Dutch physicist Heike Kamerlingh Onnes after he had been the first to succeed in liquefying helium, at the temperature of 4.2°K.[34] Attempting for 15 years to make it freeze without success, he unexpectedly found that at 2.2°K it turned into another kind of fluid, now called a superfluid, the strange properties of which were subsequently discovered by the Russian physicist Pyotr Kapitsa. Its viscosity is so low that it can flow through the finest cracks; it will creep up the walls of any container and escape through its open top; furthermore, it conducts heat 800 times better than copper. The explanation of these unusual properties of superfluid helium turned out to be that a helium atom, with its two protons, two neutrons, and two electrons — all of them fermions — is a boson and therefore subject to Bose–Einstein condensation. This is the phenomenon predicted by Einstein (mentioned in Chapter 3), when all the particles in a substance made of identical bosons congregate in their lowest energy level, their properties so correlated that they act as though they were a single entity. In other words, superfluidity is one of the rare instances in which a quintessential

---

[34] K stands for Kelvin, or degrees on the absolute temperature scale, which uses Centigrades but starts at −273°. In other words, 4.2°K equals −268.8° Celsius (−451.8° Fahrenheit).

submicroscopic quantum phenomenon becomes macroscopically observable.

Returning now to Heike Kamerlingh Onnes: having liquid helium at his disposal in his laboratory, he began to use it to cool down other materials to temperatures below 4.2°K, the way you use cold water to cool a warm pudding. Intending to study the electrical resistance of metals at low temperatures, he discovered to his astonishment that at 3°K the resistance of mercury completely disappeared. This property of *superconductivity*, he found, was shared by lead and several other metals as well. An electric current set up in a superconducting wire loop would continue to circulate forever without diminishing unless the superconductivity was destroyed by a strong magnetic field. What is more, the German physicists Walter Meissner and Robert Ochsenfeld discovered that a superconductor immersed in a magnetic field — one not strong enough to destroy the superconductivity — would expel the field completely from its interior. (Both this so-called Meissner effect and the absence of resistance have enormous potential technological applications that have not yet been fully exploited.)

It took many years for the strange phenomenon of superconductivity and the accompanying Meissner effect to be explained. The hero of that story was John Bardeen, the same man who also invented the transistor, though he did neither single-handedly. (For each of these achievements he shared a Nobel Prize in physics, the first person to win two of the prizes in science since Marie Curie won hers in physics and chemistry.) Like that of superfluidity, the fundamental explanation of superconductivity (called the BCS theory after John Bardeen, Leon Cooper, and Robert Schrieffer) rests on the formation of bosons from pairs of fermions. In this case they are "Cooper pairs" of electrons attracted to one another, in spite of their electrostatic repulsion, through a force generated by the distortion of the crystal lattice surrounding the electrons. Here again we have a macroscopically observable phenomenon caused by submicroscopic particles governed by quantum mechanics.

There are indeed no properties of solids or fluids that do not rely for their explanation on the particulate constitution of all matter. On the other hand, it is also important to recall that according to quantum field theory, particles originate as quanta of an oscillatory field. And just as the oscillations of the electromagnetic field give rise to light, so oscillations of the particles making up fluids and solids are perceived as sound. The quantization of the oscillations of electromagnetism then produces photons, the particles of light. In the same manner, the quantization of the oscillations of the constituents of solids and fluids produce quanta that are called *phonons,* the particles of sound. These particles can exist, of course, only inside solid or fluid matter; they have no independent existence in vacuum. Are they "real"? As far as physicists are concerned, phonons are objects to be dealt with just like photons; the problem of their reality matters little to them.

9

Methods and Underpinnings

In the 1950s, a great proliferation of newly discovered kinds of particles burst upon the scene, presenting physicists with a cornucopia. However, before describing the discovery of the many new kinds of particles we should understand the experimental methods used to detect the presence of these particles; then, how a novel kind of powerful explanatory mathematical method entered into physics with the ability to transform apparent chaos into an orderly scheme.

Experimental Methods

There were two fundamentally different ways in which experimental physicists working at the large accelerators gathered evidence for the existence of particles. One was by the use of detectors consisting of cloud chambers or other devices that either made the tracks of charged particles visible to the eye or to cameras or else made their presence known to a computer. These other devices were the bubble chamber, in which the particles passing through a liquid made their path known by leaving a trail consisting of tiny bubbles, or thick photographic emulsions, or spark chambers, in which charged particles passing through an electrical grid set off sparks to show their presence. Useful as all these detecting devices were, they failed to

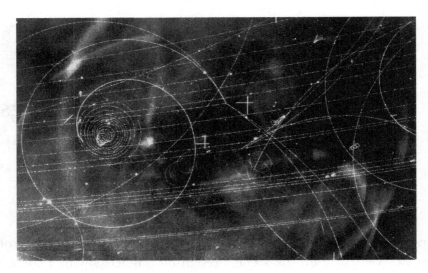

Figure 13. Traces of electrically charged particles in a hydrogen bubble chamber in a magnetic field. The almost straight tracks are those of fast particles; the curved tracks, bent by the magnetic field, are those of slower particles such as electrons or pions produced in collisions. (Photograph courtesy of the high-energy physics group at Indiana University.)

work for unstable particles which moved slowly and had such a short half-life that they would decay before they had time to leave a track.

The second mode of detection was based on a characteristic quantum mechanical phenomenon that needs some explaining. When a particle beam from an accelerator is directed at a fixed target or at another particle beam, the fraction that is scattered is called the scattering cross-section, and this number varies as the beam energy is changed. The result is plotted as a curve, tracing the scattering cross-section as a function of energy. If the two particles — one from the beam, the other from the target or from a second beam — can form a temporary union, so to speak, at a certain energy, at which they stay together for a while as a new entity before separating, then the scattering cross-section is enhanced, forming a bump in the curve called a resonance. The width of this resonance

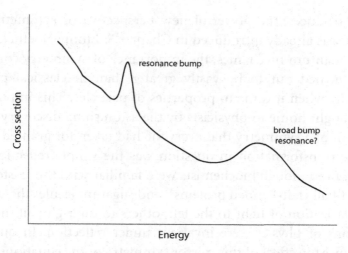

Figure 14.   A sample cross section plot. (The smooth curve is an idealized average of experimental points.)

bump is inversely proportional to the half-life of the unstable union formed: the longer it can stay together, the sharper the bump in the cross-section curve. Therefore, if the two colliding particles produce a new, unstable particle, then their scattering cross-section is expected to show an enhancement centered at the energy that corresponds to its mass by $E = mc^2$, and the width of the bump indicates its half-life. This is the way many of the new, unstable particles were discovered (see Figure 14). Of course, this method is not foolproof, especially if the unstable particle has a very short half-life. A broad enhancement in a curve is not always unambiguously recognizable. How can we tell whether it is just a normal increase in the cross-section or the result of the production of a short-lived new entity? There have been some occasions when a hastily announced discovery had to be rescinded because a bump in a plot had been identified erroneously.

A New Theoretical Tool

What was the new theoretical lens through which physicists looked at the bewildering variety of new particles they discovered? The person who introduced the powerful new perspective of symmetry into physics was already introduced in Chapter 5: Emmy Noether.

In quantum mechanics, the importance of symmetry considerations turned out to be vastly greater than in classical physics, especially when it came to properties of particles. This importance was brought home to physicists by the astonishing discovery of the violation of a symmetry that everyone had taken for granted.

The transformation in question was the simple reflection in a mirror. Even though biochemists were familiar with the existence of preferentially left-handed proteins[35] and sugar molecules that turned the polarization of light to the left, others to the right, all the basic equations of physics were invariant under reflection. In quantum mechanics the effect of this mirror symmetry of the equations is that their solutions can be classified according to their "parity." Some have parity +1, others −1; the former remain unchanged when seen in a mirror, the latter change sign. A similar situation exists for particles, the quanta arising from mirror-symmetric field equations: some have parity +1, others −1, and parity is a conserved quantum number. The product of the participating particles' parity is not allowed to change in a reaction, even if the number of particles changes.

Now, among the strange particles discovered around 1955, there were two, called tau and theta, that appeared to be identical in all respects, except that they had to have opposite parity because the parity of the initial reactants producing the tau was the opposite of the parity of those producing the theta. This "tau-theta puzzle" baffled everyone until the two Chinese-born American physicists C.N. Yang and T.D. Lee solved it by proposing simply that the two apparently different particles were really one and the same, but that in their production parity was not conserved. Of course, if that had been all, their suggestion would have been ignored; conservation of parity, after all, was considered an extremely well-established fundamental

---

[35] The mirror image of your right hand is a left hand. "Handedness" is the characteristic of an object that lacks mirror symmetry.

principle. But they went on to suggest that, since the tau and theta were produced in weak interactions, other weak interactions, beta radioactivity for example, should be carefully checked to see if there too parity conservation might be violated. They specifically pointed to one particular case that seemed never to have been tested, and when that experiment was quickly performed, another Chinese-American physicist, C.S. Wu, and her collaborators, found the Yang–Lee proposal indeed to be correct: contrary to everyone's firm conviction, the weak interaction giving rise to beta radioactivity violated the sacred parity-conservation law; in fact it strongly violated it. Nature was not mirror symmetric after all.

The newly discovered strong violation of parity conservation in beta decay had an immediate impact on the theory of the nature of the neutrino and its weak interaction with nucleons and electrons. It looked as though the neutrino was intrinsically left-handed: its spin, together with its direction of motion, resembled a left-handed screw. If its mass were exactly zero, so that it always moved at the speed of light and could never be seen at rest, there was nothing wrong with such a theory.[36]

Violations of some intuitively appealing symmetry principles notwithstanding, many interactions were found to obey new symmetries, with important explanatory consequences. That other interactions weakly broke these symmetries would have explanatory consequences of its own. Physicists suddenly had to learn a branch of mathematics dealing with symmetries called group theory, which had been quite unfamiliar to them.

---

[36] Later, however, it turned out that the neutrino masses could not be exactly zero. In order to account for the observed number of neutrinos coming from the sun, emitted by the two kinds of processes proposed by Hans Bethe to keep the sun shining, it was necessary to assume that the electron neutrino and the muon neutrino performed an oscillatory dance which transformed one into the other, and that was possible only if the masses of these neutrinos were nonzero. But this might be a result of the "Higgs mechanism."

Group theory had been invented in the early 19th century by the young French mathematician Evariste Galois just before his death in a duel at the age of 20. It was one of the many areas of mathematics developed for purposes totally unrelated to their eventual application in physics. Here is why, more than a hundred years after Galois's death, the theory he originated became a most valuable tool in quantum mechanics.

The transformations, such as translations or rotations, of a given kind of symmetry usually form a group in the mathematical sense.[37] Every given group has a certain number of "representations," each with a unique "dimension," which is easily calculated from the properties of the group. If a system is known to be invariant under all the transformations in the group, the possible "degeneracies" of its energy levels are equal to these dimensions. (A quantum-mechanical energy level is called degenerate if more than one state of the same system has the same energy; if its degeneracy is five-fold, five different states, with different characteristics, of the same system have the same energy.) The possible degeneracies of the energy levels of a system known to be invariant under a given group of transformations can therefore be determined without knowing other details about its internal interactions and can be calculated without ever having to solve the complicated equations of motion. Since particles originate as the quanta of a field, their masses are directly connected by Einstein's equation $E = mc^2$ to the quantum-mechanical energy levels of that field. The degeneracy of a level therefore tells us directly in how many different versions the corresponding particle will exist, all with the same mass. What is more, group theory also

---

[37] Here are the conditions for a collection of transformations to form a group: If the transformations S and T are in the group, so is the product ST, which means performing first T and then S (ST is not necessarily the same as TS, as you can readily see, for example, if S and T are rotations about different axes in space); if S is in the group, so is its inverse, which undoes the transformation; finally, the identity transformation, which means doing nothing, is also a member of the group.

allows us to extract information about the transition probabilities from one level to another if the system is perturbed so that the symmetry is broken, and especially to identify what kinds of transitions are "forbidden." For the corresponding particles, this means information about decay probabilities into lighter particles if they interact with others, and the absence of certain specific decays.

For example, since every particle is presumably the quantum of a field whose equations are invariant under rotations, it must be associated with one of the representations of the rotation group. These representations have dimensions 1, 2, 3, etc., which implies that the corresponding particles have spin 0, $h/4\pi$, $h/2\pi$, etc. (If a particle has spin $nh/2\pi$ then the projection of its spin on any axis can have $2n + 1$ values; it has therefore $m = 2n + 1$ different possible states. The representation of dimension m therefore corresponds to a spin value of $(m - 1)/2\ h/2\pi$.) Similarly, if we postulate that the field equation of the nucleon is invariant under rotations in a non-physical, fictitious three-dimensional space, then its quanta associated with the two-dimensional representation of that rotation group come in two versions: "isotopic spin" up and down, conventionally called neutrons and protons. (The isotopic spin has nothing to do with ordinary spin; it is a purely mathematical device.)

If all of this strikes you at this point as a meaningless mathematical game, wait until you see its organizing power when applied to the confusing variety of particles discovered during the second half of the 20th century. The electroweak theory was already an example of this organizing power. Its fusion of the weak interactions with electromagnetism had been constructed by means of a generalization, proposed by C.N. Yang together with the American physicist Robert Mills, of an abstract symmetry of electromagnetism known as "gauge invariance." From then on, this generalized symmetry was adopted by all future field theories — they were known as "Yang–Mills theories" — and the new particles were their quanta.

Strange Particles

The cornucopia of newly discovered particles started with evidence, both in cosmic-ray showers and in collisions with accelerator beams, of "strange" particles. All unstable, they were regarded as strange because, although found in relative abundance, they had long enough half-lives to leave lengthy tracks in cloud chambers and photographic emulsions. Since the production of new particles is the result of interactions between other particles when they collide, copious production is evidence that this interaction is strong. But that same interaction also makes the new particles decay, and the stronger it is, the shorter the half-life of the new entity is expected to be. The combination of abundant production and long half-lives was consequently regarded as very strange. The proposed explanation was "associated production," in which the "strange" particles were always produced in pairs, the members of which had certain properties called quantum numbers, the sum of which had to be conserved. (Think of the way a photon may, under certain circumstances, produce an electron, but only in conjunction with a positron, so as to conserve the total electric charge.) This conservation law then prevented the new particles from decaying; the explanation of the decay process had to be sought in another, weaker interaction, perhaps in beta decay, that violated the conservation law.

The newly found particles came in two kinds: the first, called hyperons, were fermions heavier than neutrons and protons, and the second, called kaons, were bosons heavier than pions but lighter than protons. The individual hyperons were named Lambda, which was neutral, Sigma, which came in positive, neutral, and negative forms, and Xi, negative and neutral; the kaons were found in positive, negative, and neutral form. A scheme of "strangeness quantum numbers," devised by the American physicist Murray Gell-Mann and the Japanese Kazuhiko Nishijima, fleshed out in detail the idea of associated production; what is more, the prediction implied by the scheme that there should be two different kinds of neutral kaons with different half-lives was soon confirmed by experiments. In addition to the light fermions, i.e., the electron, the muon, and the two different kinds of neutrino (all these weakly interacting particles were named "leptons") and the heavy bosons involved in the weak interactions leading to beta radioactivity, i.e., the W^+, the W^-, and the Z^0, the following particles were now known to exist: eight heavy fermions, i.e., the two nucleons and the six hyperons (the Lambda, the two Xis, and the three Sigmas) — these eight were called baryons — and the seven mesons (the three pions, positive, negative, and neutral, and the four kaons, positive, negative, and two different neutral ones). In 1962, another neutral meson was added to the list by the discovery of a fairly sharp resonance in collisions of positive pions with deuterons[38]; called eta, it had a mass of about 550 MeV/c^2. The situation was beginning to look more and more confusing, with particles upon particles and no underlying schema. Although the new "strangeness" quantum number introduced by Gell-Mann and Nishijima accounted well for the observed "associated productions," there was a distinct lack of rationality underlying it all. No symmetry was known that would explain the "conservation of strangeness."

The required new symmetry group was introduced in 1961 by Gell-Mann and independently by the Israeli physicist Yuval Ne'eman. Called the "eightfold way" by Gell-Mann, the technical name of the group was SU(3) and it not only had an eight-dimensional representation that exactly accommodated the eight baryons with all their quantum numbers and charges, but it also suited the eight mesons, including the newly discovered eta. However, if the fields that produced the baryons and the mesons as quanta had SU(3) symmetry, the eight baryons should all have the same mass, and so should the eight mesons. In reality these masses were not all exactly the same but differed by relatively small amounts. To take these experimental

---

[38] The deuteron is the nucleus of heavy hydrogen, called deuterium. It consists of a proton and a neutron.

facts into account, Gell-Mann and the Japanese-born American physicist Susumu Okubo suggested that the SU(3)-symmetry of the underlying field was not quite perfect; assuming it was slightly "broken" in a certain way, they could approximately calculate the resulting differences in the particles' masses, obtaining reasonable agreement with the data.

If the "eightfold way" together with the Gell-Mann–Okubo mass formula did not convince everybody that SU(3) was nature's almost perfect symmetry, then a new experimental fact finally did. It was quite analogous to when Mendeleyev introduced his periodic table of the elements, which had a number of empty slots in it. The discovery of the missing elements with the predicted chemical properties did much to convince scientists that the proposed table was right. For some psychological reason, the verification of such predictions appears to be more persuasive than the theoretical "post-diction" of already known facts. In this case, there were nine known particles that fit perfectly into a 10-dimensional representation of SU(3), with all the correct properties, but one of the slots, which should have been occupied by a negatively charged particle, remained empty. If ever found, it was to be named omega-minus; even its mass was approximately predicted by the Gell-Mann–Okubo formula. After two years of searching, Gell-Mann's predicted omega-minus was triumphantly discovered in a bubble-chamber photograph at Brookhaven National Laboratory. There could no longer be any doubt that SU(3) was a symmetry of nature, albeit slightly imperfect.

What is more, SU(3) led to a much more systematic organization of all the particles involved in the strong interactions, the so-called hadrons. In order to set up a simple organizing principle for all the baryons and mesons, Gell-Mann proposed in 1964 to use the "fundamental" representation of SU(3), which was three-dimensional, to form a new system of all the particles he and Ne'eman had previously schematized. This system would make it appear as though they could all be mathematically constructed out of three fundamental entities he called *quarks* (from a line in James Joyce's *Finnegans Wake*; rather than showing off their education in Latin as in the past, some physicists now paraded their knowledge of literature). The Russian-born American physicist George Zweig had proposed a similar scheme out of "aces," but it was Gell-Mann's name "quark" that stuck. Initially proposed as a purely mathematical book-keeping device, the quarks soon acquired reality when scattering experiments showed evidence of a certain graininess inside nucleons, reminiscent of the way Rutherford had found his fly in the cathedral, the atomic nucleus.

The principal reason why Gell-Mann was initially reluctant to regard his quarks as more than a mathematical fiction was that their electric charges had to be fractions of the electron's: two of the quarks, called "up" and "strange," had to have a positive charge 2/3 as strong as that of the electron, and the third, called "down," had to be negative, 1/3 as strong as the electron. Each baryon was to be made up of three quarks and each meson (including some newly discovered ones in addition to the pions and kaons) was to consist of a quark and an anti-quark. For example, the proton would consist of two up quarks and one down quark, making its charge 2/3 + 2/3 − 1/3 = 1 electronic charge, while the neutron would be made up of one up and two down quarks, so that its charge comes out to be 2/3 − 1/3 − 1/3 = 0. The hyperons would all contain strange quarks: those containing a strange quark or anti-quark were strange particles. The virtue of the new scheme was that it explained why only the eight-dimensional and the ten-dimensional representation of SU(3) were realized in nature, even though SU(3) had infinitely many others. The downside was that nobody had ever seen any of these particles with fractional charges and no one could find any of them in spite of diligent searches. To this day, they have never been seen outside the confines of baryons or mesons. The explanation of this mysterious absence of quarks in the wild is thought to be — though possibly the last word on this has not yet been said — that the attractive force between them, which keeps them together inside baryons and mesons, is relatively weak at small distances but remains strong and undiminished at large distances. The intuitive argument then says that any attempt to separate two quarks requires enough energy to produce a new quark–anti-quark pair, thereby frustrating the separation effort.

Attempts at bringing order to the welter of new particles by means of quarks were further complicated when in 1975 Martin Perl experimentally discovered yet another weakly interacting particle. He called it tau; though apparently a lepton, it weighed more than twice as much as the proton. Just like the electron and the muon, it came with its own neutrino. This increased the total number of leptons to six: electron, muon, tau, and their three different neutrinos. In order for the underlying field theory to be renormalizable, i.e., to be free of nonsensical infinities, it turned out that the strongly interacting particles, the hadrons, had to be organized in parallel with the weakly interacting ones, the leptons. Specifically, there had to be as many quarks as there were leptons, namely six. The "flavors" of these six quarks were called *up, down, strange, charmed, top,* and *bottom.* What is more, in order to reconcile forming all the needed hadron states out of quarks with Pauli's exclusion principle, each of the six quarks had to come in three "colors" (a purely metaphorical name with no relation to real color).

As we have seen, in order to rationalize the zoo of newly discovered particles it was found necessary to postulate the existence of a sizeable number of additional ones. Experimentalists had to search for years to find many of them, but eventually they did. The first "charmed" meson was found in 1976; it came both in neutral form and positively charged, the former made up of a charmed and an anti-up quark and the latter composed of a charmed and an anti-down quark. The top quark was not found until 1995 — of course only indirectly, never in the wild — and was almost as heavy as a tungsten atom. The first "charmonium," made up of a charmed quark and a charmed anti-quark, was found simultaneously by

Samuel Ting at Brookhaven National Laboratory, who called it "J," and Burton Richter at the SLAC laboratory at Stanford University, who called it "Psi." To keep the peace it was named "J/Psi." Confusing as particle physics seemed for a number of years, the idea that all the known "elementary" particles were made up of more fundamental constituents called quarks, which came in six "flavors" and three "colors," made order out of what appeared like pure chaos, just as the Rutherford–Bohr picture of the atom organized the chemical table of the elements.

The Standard Model

But where was the theory underlying this conglomeration of particles? There had to be a quantum field theory that would produce the quarks as its quanta. Modeled on the enormously successful QED, the new theory was called quantum chromodynamics, or QCD (the chromo prefix because of the "color" of the quarks). Instead of the positive and negative electric charges of QED, it had three "colors," and in place of photons it had "gluons." An important difference from QED was the fact that, whereas photons do not directly interact with one another, gluons do attract each other. In fact they should form "glueballs," though none have been found as yet. Another important difference between QED and QCD is that, whereas QED led to a number of calculations that were experimentally verified with astonishing precision, QCD does not lend itself to any such verifiable calculations; its equations are too difficult to crack and solutions not easily approximated. Nevertheless, the scheme of particles it produces — the quarks and their bound states, the baryons and mesons — is so successful that it forms a firm foundation for this quantum field theory. The combination of QCD as the theory of the strong interactions, together with the electroweak theory melding electromagnetism with the weak interactions, is called the Standard Model of all the known particles in the world.

There is one other particle whose theoretical existence arises from symmetry considerations, and one new symmetry that implies the existence of more particles. In order for the electroweak field theory to be renormalizable, its bosons, the Z and Ws, had to be assumed to have zero mass, like the photon. Their observed, nonzero masses are assumed to be generated by a mathematical mechanism found by the British physicist Peter Higgs, called dynamical symmetry breaking. Similarly for the masses of all the quarks, which in a "grand unified" version of the Standard Model also must vanish; their real values have to be theoretically produced by the Higgs feature. If this mechanism is in play, however, it should generate a particle of its own, the mass of which is hard to predict except that it is expected to be heavy. The Higgs particle remains the grail of particle physics, with enormous new accelerators slated to hunt for it, both at CERN in Geneva, Switzerland, and at Fermilab near Chicago. It has not yet been found.

The new postulated symmetry, called "supersymmetry," is an invariance under a transformation that connects bosons with fermions. Supersymmetric theories automatically predict the existence of boson partners for all fermions and fermion partners for all bosons. In other words, in addition to all the quarks, which are fermions with spin $h/4\pi$, there should be as many "squarks," which are bosons with spin 0; in addition to photons with spin $h/2\pi$, there should be "photinos" with spin $h/4\pi$. This scheme, however, is entirely conjectural at this point, and there is no evidence for the existence of either photinos or squarks.

Are we now at the end of the search for all the particles that make up the world? There is reason to think so. The quark scheme appears to be closed, with no other colors or flavors to be added. Much will depend on whether the elusive Higgs particle is ever found. If neither of the new accelerators at CERN and Fermilab is able to discover it, the whole theoretical scheme constructed as the Standard Model of particle physics is likely to collapse. If it is found, there will be order in our understanding of the way the world is made up of particles. And all of this order is ultimately based on quantum field theory.[39]

However, the cloud casting a dark shadow on theoretical physics is the fact that no version of the Standard Model has been reconciled with Einstein's General Theory of Relativity. We have one theory that deals with motion on a large scale and another that explains submicroscopic structure on the basis of abstract symmetry principles, splitting the world into two spheres analogous to Aristotle's way. It is this theoretical gulf between two theories, each individually well-confirmed within its own domain of applicability, that has

---

[39] We may still ask, how did our universe become filled with particles? In Chapter 7 we saw that our understanding of nuclear physics was able to account for the existence and distribution of all the elements in the world by nucleosynthesis, but the calculation by Fowler and Hoyle started with the existence of baryons as well as photons at an early stage of the universe. Presumably, initially, right after the beginning of the universe, at the Big Bang, there was only "pure energy" and no matter. This means there existed nothing but electromagnetic radiation at extremely high temperature, the cooled remnants of which were observed for the first time in 1965 by Arno Penzias and Robert Wilson. At this early stage of the universe, matter in the form of baryons, or clumps of quarks and anti-quarks, was generated from photons, as expressed by Einstein's $m = E/c^2$. In other words, the first particles with rest mass, the defining characteristic of matter, made their appearance in the world. This was the beginning of the material universe. And all of it was made of particles rather than antiparticles! This is still one of the major mysteries left for 21st century physics to solve. The problem arises from the fact that one of the important conservation laws embedded in the Standard Model is the conservation of baryons, which implies that no baryon can be created out of "pure energy" without at the same time making an antibaryon. But where are all those antibaryons? Could there be regions of the universe made up entirely of antiparticles? Antistars and antigalaxies would look the same to us as stars and galaxies, so how could we tell? The trouble is that if ever a star or galaxy, or even an interstellar gas cloud, came near an antistar, antigalaxy, or antigas cloud, they would annihilate one another in a spectacular explosion ending up in nothing but radiation. Astronomers have never observed such horrendous events. It is therefore unlikely that the universe is made up equally of baryons and antibaryons, and we are left with a deep puzzle.

given rise to the excitement generated by various kinds of "string theories" and their supersymmetric versions called "superstring theories." These new theories manage to meld general relativity with our systematic knowledge of particles in the world, but at a price: here particles are thought to be "really" made up of higher-dimensional vibrating strings. None of these theories has yet yielded any predictions capable of experimental verification, nor are they likely to do so in the foreseeable future. Nevertheless, there are many physicists who are devoting their lives to working on variants of these theories, which may all turn out to be chimeras. They cling to the hope that their theory may turn out to be the analog of Bohr's model of the atom — albeit without the analogous experimental evidence — to be followed by later justification.

At this stage of physics at the beginning of the 21st century it looks as though the visible world consists entirely of the particles systematized by the Standard Model, all ultimately based on quantum mechanics in the form of quantum field theory. (That there also appears to be an invisible world made up of "dark matter," comprising most of its mass, is a puzzle left for 21st century physicists to solve.) The large-scale motions of the universe, on the other hand, seem to be well described by Einstein's General Theory of Relativity. The implied division of the world may not please us, but for the time being it is the best we can do.

Epilogue

Our survey has clearly shown that quantum mechanics and quantum field theory achieved their greatest accomplishments in explaining the particulate structure of the world, an area in which philosophical speculation had reigned for more than 1,500 years and that finally made contact with scientific reality by means of classical physics and chemistry. The achievements of quantum field theory were admirable indeed. In spite of the fact that quantum mechanics intrinsically deals with probabilities rather than certainties, its mathematical techniques have led to numerical predictions of data that were experimentally verified, sometimes with an unprecedented accuracy of better than one part in a billion. In many other instances, quantum field theory has been able to exploit very general, abstract symmetry postulates to arrive at qualitative and quantitative classifications and clarifications of a bewildering mass of observational discoveries. Einstein's correct characterization of quantum mechanics as not dealing directly with reality notwithstanding, the accomplishments of the theory in the area of structure, initiated by both Einstein and Bohr, were surely as close to reality as even Einstein could have wished.

On the other hand, when it comes to the explication of motion, quantum mechanics has been less successful. In the first place it was handicapped by its probabilistic character, even though, from a practical point of view this did not fundamentally distinguish it from classical mechanics, which, as we have seen, implied unpredictable chaos for most mechanical systems. But it was in this area that the divorce of theory from a direct contact with reality became most manifest. And there is no way of avoiding the fact that, as a matter of principle, quantum mechanics differed fundamentally from classical physics in not dealing directly with reality itself, just as Einstein had insisted.

While our survey has shown that the distancing of the equations of motion from real, observable movements of objects in the everyday world had been, historically, a gradual process since their first formulation by Aristotle, quantum mechanics represented a definite break. The Newtonian equations of motion and their radical modifications by Einstein in his special and general theories of relativity were non-realistic only in the sense of being abstract and unintuitive; they still directly described at least in part the reality of physical systems. In generating his path-breaking relativity theories Einstein prided himself in his insight into reality as he saw it, and, after some hesitation, scientists almost universally agreed and praised him for it. The equations of motion of quantum mechanics, on the other hand, made no pretense of being mathematical descriptions of how physical systems themselves moved; they confined themselves to the predictions of the probabilities of the outcomes of experimental measurements and how these changed in the course of time. This was the fundamental point to which Einstein objected and that Bohr accepted with his words, "It is wrong to think that the task of physics is to find out how nature *is*. Physics concerns what we can *say* about nature." If all we can say about certain aspects of nature is the outcome of experimental test, then so be it! Physicists have no business asking for more if nature refuses to give it.

The fact that since the beginning of the 20th century the focus of submicroscopic physics has shifted away from describing and explaining motion[40] to explaining structure, which quantum mechanics did superbly well, has removed most of the power of Einstein's critique. There can be no question that Bohr has won the game. A radical modification of quantum mechanics is not among the pressing problems facing the physicists of the 21st century, and there is no indication that Einstein's lead will be followed.

---

[40] This is not the case, however, in all branches of physics. But those fields like fluid mechanics and cosmology, in which motion is still at the center of attention, make little or no use of quantum mechanics for the explication of this motion.

References and Further Reading

Amaldi, Ginestra, *The Nature of Matter: Physical Theory from Thales to Fermi.* The University of Chicago Press, 1966.

Anstey, Peter R., *The Philosophy of Robert Boyle.* New York: Routledge, 2000.

Blackmore, John T., *Ernst Mach: His Life, Work, and Influence.* Berkeley: University of California Press, 1972.

Bohr, Niels, *Atomic Theory and the Description of Nature.* Cambridge University Press, 1961.

Broda, E., *The Intellectual Quadrangle: Mach — Boltzmann — Planck — Einstein.* Geneva: CERN, 1981.

Brown, L.M., A. Pais and B. Pippard, *Twentieth Century Physics.* Bristol: IOP Publishing, Ltd., 1995.

Brush, Stephen G., *The Kind of Motion We Call Heat: A History of the Kinetic Theory of Gases in the 19th Century,* Book 1. Amsterdam: North Holland, 1976.

Cercignani, Carlo, *Ludwig Boltzmann: The Man Who Trusted Atoms.* Oxford University Press, 1998.

Einstein, Albert, *The World as I See It.* New York: Covici, Friede, 1934.

Einstein, Albert, *Ideas and Opinions.* New York: Crown Publishers, 1954.

Fermi, Laura, *Atoms in the Family.* London: Allen and Unwin, 1955.

Glashow, Sheldon, *Interactions: A Journey Through the Mind of a Particle Physicist and the Matter of this World.* New York: Warner Books, 1988.

Gregory, Joshua C., *A Short History of Atomism: From Democritus to Bohr.* London: A&C Black, Ltd., 1931.

Home, Dipankar and A. Whitaker, *Einstein's Struggles with Quantum Theory: A Reappraisal.* New York: Springer, 2007.

Hooper, Dan, *Dark Cosmos: In Search of Our Universe's Missing Mass and Energy.* New York: Harper Collins, 2006.

Isaacson, Walter, *Einstein: His Life and Universe,* New York: Simon & Schuster, 2007.

Kragh, Helge, The Vortex Atom: A Victorian Theory of Everything, *Centaurus* 2002, Vol. 44, pp. 32–114.

Kragh, Helge, *Entropic Creation.* Aldershot, Hampshire: Ashgate Publishing Ltd., 2008.

Lindley, David, *Uncertainty: Einstein, Heisenberg, Bohr, and the Struggle for the Soul of Science.* New York: Doubleday, 2007.

Newman, William R., *Atoms and Alchemy: Chymistry and the Experimental Origins of the Scientific Revolution.* University of Chicago Press, 2006.

Newton, Roger G., *What Makes Nature Tick?* Cambridge, MA: Harvard University Press, 1993.

Newton, Roger G., *Thinking about Physics.* Princeton University Press, 2000.

Newton, Roger G., *From Clockwork to Crapshoot: A History of Physics.* Cambridge, MA: Belknap Press, 2007.

Petersen, Aage, *Quantum Physics and the Philosophical Tradition.* Cambridge, MA: MIT Press, 1968.

Reeves, Richard, *A Force of Nature: The Frontier Genius of Ernest Rutherford.* New York: W.W. Norton & Co., 2008.

Rosenblum, Bruce and F. Kuttner, *Quantum Enigma: Physics Encounters Consciousness.* Oxford University Press, 2006.

Sarton, George, *A History of Science: Ancient Science Through the Golden Age of Greece.* Cambridge: Harvard University Press, 1952.

Sarton, George, *A History of Science: Hellenistic Science and Culture in the Last Three Centuries B.C.* Cambridge: Harvard University Press, 1959.

Schlosshauer, Maximilian, *Decoherence and the Quantum-to-Classical Transition,* Berlin: Springer, 2007.

Segrè, Gino, *Faust in Copenhagen: A Struggle for the Soul of Physics.* New York: Viking, 2007.

Selleri, Franco, *Quantum Paradoxes and Physical Reality.* Dordrecht: Kluwer Academic Publishers, 1990.

Shapin, Steven and S. Schaffer, *Leviathan and the Air-Pump: Hobbes, Boyle, and the Experimental Life.* Princeton University Press, 1985.

Squires, Euan, *The Mystery of the Quantum World.* London: IOP Publishing Ltd., 1994.

Weinberg, Steven, *The First Three Minutes: A Modern View of the Origin of the Universe*. New York: Basic Books, Inc., 1977.

Westfall, Richard S., *Never at Rest: A Biography of Isaac Newton*. Cambridge University Press, 1980.

Westfall, Richard S., *The Life of Isaac Newton*. Cambridge University Press, 1993.

Index